Matter, Energy, and Life An

Matter, Energy, and Life An introduction to chemical concepts

fourth edition

JEFFREY J. W. BAKER
Wesleyan University

GARLAND E. ALLEN
Washington University

THE BENJAMIN/CUMMINGS PUBLISHING COMPANY, INC.

Menlo Park, California • Reading, Massachusetts
London • Amsterdam • Don Mills, Ontario • Sydney

Library of Congress Cataloging in Publication Data

Baker, Jeffrey J W
Matter, energy, and life.

Bibliography: p.
Includes index.
1. Biological chemistry. I. Allen, Garland E., joint author. II. Title.
QH345.B32 1980 574.19'2 80–17946
ISBN 0–201–00169–1

ISBN 0–201–00169–1
JK–DO–89

Physical and chemical approaches to problems in biology have become increasingly productive in recent years. Major advances in the understanding of life processes have been made through research in such specialties as biophysical chemistry, molecular biology, biophysics, and electrophysiology. Continuing progress will require an ever more perceptive study of the interactions of matter, energy, and information in biological systems.

From J. L. Oncley et al., *Biophysical Science, A Study Program.* New York: John Wiley and Sons, Inc., 1959. p. 1.

Preface to the Fourth Edition

The widespread acceptance of *Matter, Energy, and Life* since its first appearance 15 years ago has been most gratifying, and it is with pleasure we present our users, both new and old, with this fourth edition. As with previous editions, we have relied heavily upon user feedback, both instructor and student, to improve clarity and increase comprehension, as well as concentrating on keeping the content up-to-date and consonant with significant new developments in the fields of inorganic and organic chemistry and biochemistry. The increasing use of *Matter, Energy, and Life* in programs for health care professionals has led us to incorporate supplements designed to show students how the theoretical material they are learning relates directly or indirectly to laboratory or clinical analyses and techniques. For outstanding assistance with these we are indebted to Dr. Alice Holtz of Boston University.

The reader who is familiar with the earlier editions of *Matter, Energy, and Life* will find that the goals and organization of the book remain the same. The level of presentation has not changed, and the chapters remain as independent of each other as possible to permit selective, nonsequential use. The book remains introductory and requires no prior knowledge of atomic or molecular structure or the principles of chemical bonding.

The authors wish to thank James H. Funston, formerly Instructor of Biology at Holy Cross, Worcester, Massachusetts, for applying his expertise in both biochemistry and editorial revision. His considerable effort has helped to make the present edition a significant step toward our goal of maximum accuracy and readability.

Middletown, Connecticut J.J.W.B.
St. Louis, Missouri G.E.A.
November, 1980

Contents

Chapter 1

Matter and Energy

The first atomic bomb exploded on a New Mexico desert in 1945. The sound foretold the end of World War II and the beginning of a new scientific era. Within the bomb, a small amount of matter became a large amount of energy. Long thought by the public to be separate and distinct, matter and energy were dramatically shown to be closely interrelated. This point would be expanded, were this book for physics or chemistry students. But it is not. It is for biology students.

In the chemistry of life, very little matter is converted into energy. So little, in fact, it can be ignored. Therefore, biologists still like to make a general distinction between them. This book will deal with matter and energy as though they were both clearly defined and separate concepts. In reality, they are neither.

What are matter and energy? What role do they play in the physics and chemistry of living organisms? Neither of these questions has yet been completely answered. Nor will they be in this book. Yet both matter and energy play such an important part in the living world that some knowledge of their characteristics is essential to the study of biology. The reason is quite simple. All living things are composed of matter. The expression of the release and use of energy by living matter is the thing we call "life." When this release and use of energy ceases, we call it death.

1–1 MATTER: CONCEPTS OF MASS, WEIGHT, AND DENSITY

Matter can be defined as that which has *mass* and occupies space. Stones, water, gases, animals, or plants: all are composed of matter. Since the scientific conceptions of mass and weight are important to an understanding of matter, it is well to examine them at the outset.

When an object is placed on a scale, it registers a certain *weight*. This means that the force of gravity pulls this object toward the earth to a degree measured by the lowering of the scale's platform. The greater the attraction between the earth and the object, the greater the weight of the object. Weight can be measured in such units as ounces, pounds, or tons. So long as the same scale of measurement is used, the relative weights of objects may be easily established and compared.

However, the use of weight as a concept predates its use in science. Therefore it has very serious limitations for modern research. A man in space may encounter a situation where neither he nor the objects around him seem to have any weight. Objects within the spacecraft not fastened down float about freely in their condition of so-called "weightlessness."

The reason is simple. Weight varies according to the force of gravitational attraction. Gravitational attraction, in turn, is dependent upon several factors. One of these is the distance from the center of the earth. A man who weighs 144 pounds on the earth's surface would weigh only 36 pounds when he was 8000 miles high, 16 pounds at 12,000 miles, and 9 pounds at 16,000 miles. Gravity, and therefore weight, is dependent on another factor. A 100-pound man would weigh only 20 pounds on the moon and about 37 pounds on Mars. Here the influencing factor is the quantity of matter in the body responsible for the pull of gravity. *The weight which a given bit of matter possesses is directly related to the force of gravity at the place where the measurement is made.*

The scientist, on the other hand, wants to describe matter in terms which are independent of the object's location. Certainly objects in a spaceship do not change their physical and chemical characteristics in the weightless condition. They still exist, occupy the same amount of space, and look the same as they do on earth. The weight of the objects has changed, but the quantity of matter present is still the same. The measure of this quantity of matter in the objects is called its *mass*. The *gram* is the fundamental unit of mass.

1–1
The scale on the left can determine weight but not mass. High above the earth, a reading would be less than it would deep in a valley. The balance scale on the right can determine both mass and weight. Since the force of gravity is the same on both pans, its effect is cancelled. Mass can therefore be determined by comparison with objects whose masses are known.

The mass of an object is not necessarily related to its size. A lead ball one foot in diameter will, of course, have less mass than a lead ball two feet in diameter. However, a lead ball one foot in diameter has a good deal more mass than an aluminum ball of the same dimensions. It is apparent, then, that there is another factor influencing mass. This factor is the *density* of matter. Density is a measure of the amount of matter in a given volume of space. Thus, the greater the amount of mass in a certain space, the greater the density. Cities, for example, are more densely populated than rural areas. This means that in equal areas of land, there are more people in the city than in the country.

Consider two solid aluminum spheres. The first is one foot in diameter, the second ten feet in diameter. Obviously, the mass of the ten-foot sphere is greater. The density, however, is the same for both.

1–2 ENERGY

In contrast to matter, energy neither occupies space nor has weight. There is no such thing, for example, as "3 cubic feet" or "4 pounds 3 ounces" of energy. How, then, can an amount of energy be measured?

Energy can only be measured by its effects upon matter. In general, the greater the effect, the greater the amount of energy. For example, under similar conditions, a stick of dynamite causes more damage to a house (matter) than a firecracker does. The dynamite releases more energy. The distance a ball (matter) travels after being batted is often, though not always, related to the amount of energy exerted by the batter.

The more matter there is to be moved, the more energy must be supplied to move it. In moving equal distances, a large animal generally uses more energy than a small one. This is one reason why large animals are usually less active than small ones. It also helps to explain why an overweight man places greater strain on his heart than a man of normal weight.

It is often useful to speak of energy as existing in one of two kinds, *kinetic* or *potential*. *Potential energy is energy that is stored, or inactive.* A stick of dynamite represents a great deal of potential energy. When released, potential energy is capable of causing an effect on matter. However, when it does so it is no longer potential energy. It is kinetic energy.

Kinetic energy is energy of motion. It is in the process of causing an effect on matter. Kinetic energy can be measured by determining how much matter it moves in a given period of time, and how far and how fast it moves it. Imagine a boulder perched on top of a hill, ready to roll down if given a slight push. In this state, the boulder and the hill are a physical system in which there is a certain amount of potential energy. As the boulder rolls down the hill, the potential energy in the system becomes kinetic. Matter is being moved. When the boulder reaches the bottom, it has converted all of its potential energy into kinetic energy. If the original system is to be re-established, the boulder must be moved back to the top of the hill. This requires an expenditure of energy. The energy necessary to get the boulder up the hill again is the same as the potential energy present in the system at the start. The conversion of potential energy to kinetic energy (the boulder going down the hill) and the conversion of kinetic energy to potential (moving the boulder back up the hill) has thus taken place.

Energy is constantly being changed from kinetic to potential and back again. The biological world is no exception. Living organisms operate by changing the potential energy found in foodstuffs into the kinetic energy of muscle contraction, manufacture of needed structural parts, etc. Energy is independent of life. *Life, however, is completely dependent upon energy.*

1–3 THE FORMS OF ENERGY

In addition to the two kinds of energy (potential and kinetic), five forms are recognized. These are *chemical, electrical, mechanical, radiant,* and *atomic energy*. The last of these, atomic energy, has little direct relationship to the normal functioning of living organisms. For this reason, atomic energy will not be included in this discussion. All forms of energy exist as either kinetic or potential energy.

Chemical energy. Chemical energy is the energy possessed by chemical compounds. Gasoline, for example, is basically a mixture of compounds that possess a certain amount of potential chemical energy. When gasoline is burned, its potential chemical energy is converted into kinetic energy. In the automobile engine, this kinetic energy moves the pistons up and down.

By releasing its chemical energy, gasoline is changed into two less complex substances, oxidized carbon and water. Like the boulder at the top of the hill, the gasoline contained a certain amount of potential energy. After the burning reaction, the end products resemble the boulder at the bottom of the hill. In each case, potential energy has been converted into kinetic.

Chemical energy can also be described in terms of building up chemical compounds from simpler parts. For example sunlight (radiant energy) is used in the leaves of green plants to build sugar and other compounds from carbon dioxide and water. This involves many chemical reactions. The end result is potential chemical energy stored in the sugar molecules. In this case, chemical energy is being *stored* by chemical reactions. This is analogous to carrying the boulder to the top of the hill.

Chemical compounds vary widely in both the amount of potential chemical energy they contain, and the ease with which this energy can be released. Gasoline, sugars, and fats contain a great deal of energy that is easily released. Other compounds such as water or carbon dioxide also contain potential chemical energy. However, a great deal of energy must be *expended* before the energy contained in these compounds is released.

Chemical energy is the most fundamental form of energy in the life processes. Every thought, every nerve impulse, every muscle movement, indeed every activity of any sort shown by living organisms is ultimately traceable to the release of chemical energy.

Electrical energy. Electrical energy is a result of the flow of electrons along a conductor. Electrons are the outermost parts of atoms. Nearly everyone has shuffled his feet across a rug, touched a metal light switch plate, and experienced a shock. There was a rapid flow of electrons from the metal plate to the individual on the rug. This was kinetic electrical energy. Potential electrical energy exists anywhere that a flow of electrons is possible.

Electrical energy is of great importance to man, mostly because it can be converted to other, more usable forms. For example, by offering resistance to a flow of electrons, heat is produced in the coils of an electric stove. By passing an electric current through a metal filament, electrical energy can be converted to light and heat, as in a light bulb or a toaster.

The movement of electrons is also important in living organisms. Electrons do not flow through cells in the same manner that they flow along a copper wire. However, electrons play a major role in the energy changes within living organisms. Electrochemical reactions, combinations of electrical and chemical energy, play a large part in the functioning of the brain and the rest of the nervous system.

Mechanical energy. Mechanical energy is energy directly involved in moving matter. The rolling of a boulder down a hill is an example of mechanical energy. Potential mechanical energy is converted to kinetic mechanical energy as the boulder moves from the top to the bottom of the hill. Simple machines, such as a lever or a pulley, are examples of devices designed to conserve mechanical energy and make its usage more efficient. By using a crowbar, for example, the energy expended in moving a heavy object is directed to the point where it will be used most efficiently.

One of the key features of living organisms is their ability to move independently. This movement involves the conversion of potential chemical energy into kinetic chemical energy, resulting in the contraction of muscles. Since in many organisms the muscles act upon the bones, which serve as levers, the total movement of such an organism demonstrates kinetic mechanical energy.

Radiant energy. Radiant energy is energy which travels in waves. Two well-known examples of radiant energy are visible light and heat. However, radiant energy also includes radio waves, infrared and ultraviolet light, x-rays, gamma, and cosmic rays. Most of these cannot be detected by our sense organs.

Radiant energy is used in less obvious ways than are some of the other forms of energy. In living organisms, radiant energy is useful in *photochemical* (*photo-*, light) reactions. A photochemical reaction occurs when light strikes the leaf of a green plant. A chemical compound in the leaf called *chlorophyll* is affected by absorbing certain wavelengths of light energy. This energy can then be converted by the plant into chemical energy and stored in food substances.

Heat energy produces its effect upon matter by speeding up the motion of the particles of which matter is composed. Heating water, for example, causes increased motion of the water molecules. If enough heat energy is supplied, boiling results. Here the water molecules are moving too fast to remain in the liquid state. They move off into the air as a gas. As we shall see, the role of heat energy in speeding up the movements of atoms and molecules has great significance to living organisms.

1–4 THE TRANSFORMATIONS OF ENERGY

All the forms of energy are interrelated and interconvertible. The conversion of one form of energy into another goes on continually. It is the basis upon which all living organisms maintain themselves. The life processes of living cells are driven by the release of chemical energy. The chemical energy may be converted into mechanical energy for motion or used to build other chemical compounds. *Energy is constantly being released and stored in living systems.*

1–2
This device, known as a radiometer, shows that light may indirectly cause matter to move. Light striking the black sides of the radiometer vanes is absorbed, and heats up this side of the metal. Light striking the silvered side is reflected. Thus this side does not heat up as much as the blackened sides. The temperature difference between the black and silvered sides causes convection movement of the air remaining in the radiometer from the cooler surface to the warmer one, building up a greater pressure on the warmer surface. Some physicists believe that the preceding explanation is not completely satisfactory. They feel that the full reason why the radiometer vanes move is yet to be explained.

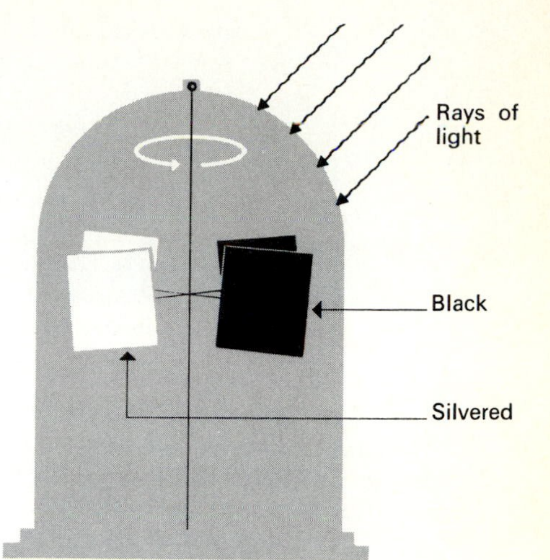

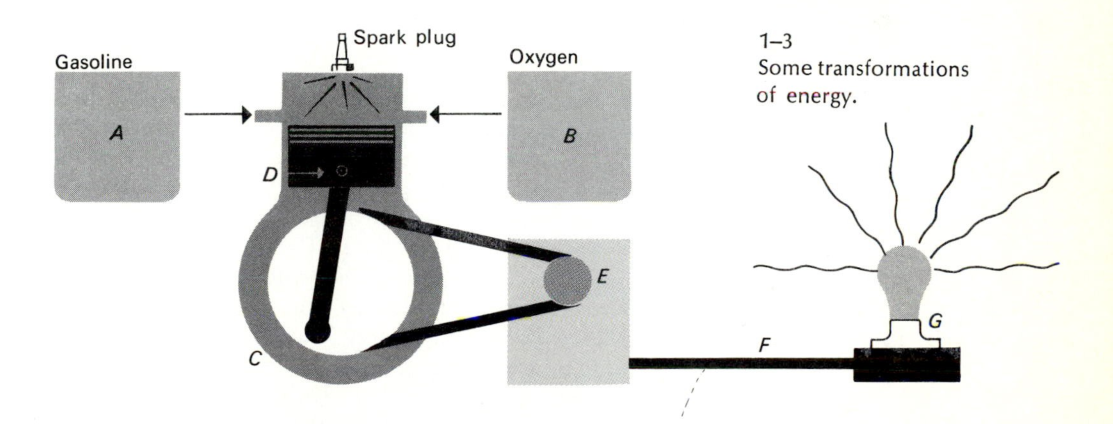

1–3
Some transformations of energy.

Figure 1–3 shows the conversion of chemical energy into radiant energy from the light bulb. The fuel, gasoline (*A*), contains a certain amount of potential chemical energy. This is released when electrical energy from the spark plug ignites the gasoline-oxygen mixture. The resulting kinetic chemical energy is transformed into heat energy. The heat energy causes increased motion of the molecules of gas within the cylinder. An increase of pressure within the cylinder results, which forces the piston (*D*) downward. A wheel attached to the generator (*E*) is rotated (mechanical energy), producing an electric current along the wire (*F*). The electric current passes through the metal filament of the bulb (*G*), where it is converted into heat and light. The light energy passes to the eye of the observer, where it starts a chemical process within cells of the retina. Thus several transformations of energy have occurred.

1–4
A biological food chain, involving several transformations of energy. Radiant energy from the sun is transformed into chemical energy during the food-making process in green plants. The cow eats the grass, and the chemical energy of plant substances is converted into the chemical energy available in milk or the cow's flesh. A man drinks the milk and further energy transformations occur. Some of this energy may be converted into mechanical energy for movement. In all of these transformations, some of the energy is lost to the system for useful work. Most of the energy transformations represented above are less than 30% efficient.

How does the amount of radiant energy emitted from the light bulb compare with the amount of potential chemical energy in the original fuel? If measurements were made under carefully controlled conditions, it would be found that a great deal of energy is *lost*. The total amount of radiant energy given off is much less than the total potential chemical energy in the gasoline. The reason? The transformation of energy is never 100 percent *efficient*. Not all of the potential energy converted into kinetic energy is used. Some of it is lost as heat into the atmosphere. Some energy is used in overcoming the friction of the moving parts of the engine and generator. Much of the light from the bulb never reaches the eye. These energy losses make the process quite inefficient. Most automobile engines are less than 30 percent efficient. This means that less than one-third of the energy released from each gallon of gasoline is actually used in the forward motion of the car.

None of the energy in these transformations is actually lost in the sense of being unaccounted for. Careful measurements of energy conversions have led to formulation of the Law of the Conservation of Energy, or the *First Law of Thermodynamics*. This law states that during ordinary chemical or physical processes, energy is neither created nor destroyed. It is only changed in form. The First Law of Thermodynamics provides a useful framework for studying energy transformation in living systems. The biologist realizes that the energy an organism derives from its food is never destroyed. It is merely lost for useful work during transformation. *The success shown by some organisms in their competition with others is often due to the higher degree of efficiency with which they use the energy available to them.*

1–5 SOME CHARACTERISTICS OF LIGHT

Light is one of the most important forms of radiant energy for living organisms. All life processes ultimately depend upon the ability of green plants to capture light energy and use it to produce food.

What is light? While it is possible to describe some of its characteristics such as color, and how it is reflected, the true nature of light is still not fully understood. Like all forms of energy, light can be described only in terms of its detectable effects upon matter.

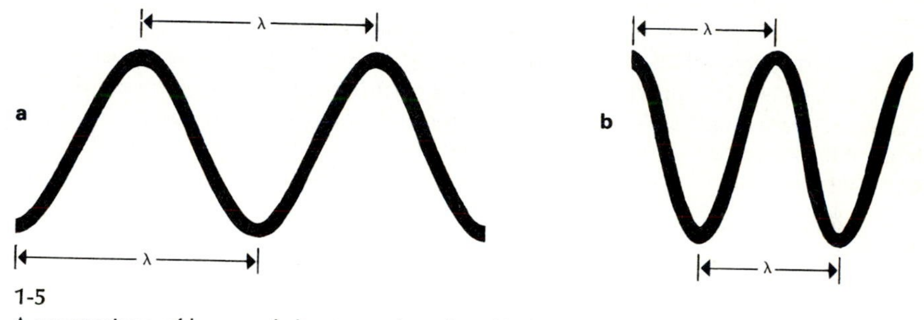

1-5
A comparison of long and short wavelengths of light. (a) red, $\lambda = 7000A$; (b) blue, $\lambda = 4000A$.

If a beam of ordinary sunlight is passed through a *prism,* the rays of light are bent. The white light is broken up into seven major colors: red, orange, yellow, green, blue, indigo, and violet. A similar effect is familiar in the rainbow. These colors make up the *visible spectrum,* so called because it is visible to most human eyes. The spectrum is continuous, with each color grading into those on either side.

Attempts to explain the way in which light is separated into these seven colors are connected with the question of how light is transmitted through space. The *wave theory* considers that light travels in a manner analogous to waves on the surface of a pond. The *quantum theory* holds that light is composed of tiny particles called *quanta* or *photons.* These particles, which have no mass, are given off by any light-emitting object and travel through space until they interact with a material object.

Both the wave theory and the quantum theory explain certain characteristics of light that the other does not explain. In this sense, although the two concepts of light are quite different, both are useful for understanding certain properties of light.

If a stone is tossed into a pond of water, waves are produced that spread out in all directions. The waves will be a certain distance apart. The distance measured from the crest of one wave to the crest of the next wave is the *wavelength.* The wavelength might be one inch, for example. Thus within a distance of one foot on the surface of the water, there will be twelve waves. If several stones are tossed into the water in rapid succession, a greater number of waves may be produced within the distance of one foot. Since there are more waves within this one-foot distance, they must have a shorter wavelength than those waves produced when just one stone was tossed into the water—perhaps only half an inch.

Thus, as the wavelength decreases, the *frequency* of waves increases. Given equal distances of one foot, there will be more short waves present than long waves. The wavelength was decreased, but the number of waves was increased. The same holds true in the wave theory of light. When the wavelength decreases, frequency increases, and vice versa.

This analogy of waves on a pond merely offers a *model* for understanding the wave theory of light and explaining some of its effects. The expression of each color of light depends upon a different wavelength. *Wavelength and frequency are two of the ways in which radiant energy can be described* (see Fig. 1–6).

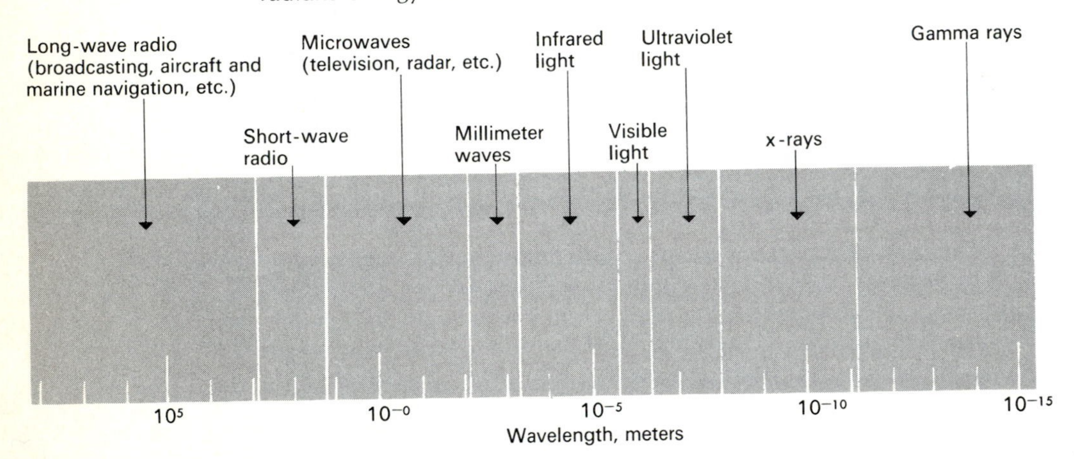

1-6
Scale representation of the complete spectrum of radiant energy. On the left are the long wavelengths of radio and microwaves. To the right are the short wavelengths of x-rays and gamma radiation. Note that the visible spectrum is only a small fraction of the whole spectrum.

The energy conveyed by a beam of light is directly related to both the wavelength and the frequency. As a matter of fact, the amount of radiant energy conveyed by a beam of light varies *directly* with the frequency. The higher the frequency the greater the energy. However, the energy varies *inversely* with the wavelength. In other words, as the wavelength of light is increased, the amount of energy it conveys decreases, and vice versa. In the visible light spectrum, red light has the longest wavelength. Violet has the shortest. The other colors fall between these two extremes. Thus, all other conditions being equal, red light conveys less energy than orange; orange less than yellow; yellow less than green; and so on. Violet light, with the shortest wavelength of visible light, conveys the most energy.

An analogy may help to make clear the relationships of energy, wavelength and frequency. Picture a man standing beside a highway along which cars are passing. Every car is traveling at the same constant speed, and all are equally spaced a thousand feet apart. The distances from the middle of one car to the middle of the next represent one wavelength of light. As the man counts the cars, he finds that the number passing each minute, or their *frequency,* is constant. Assume that 15 cars pass each minute. If the velocity remains constant, but the traffic becomes twice as heavy, the distance between the cars is reduced by one-half. Therefore, the frequency with which the cars pass in one minute is now increased to 30. So it is with light waves; the shorter the wavelength, the greater the frequency.

Now suppose that each car represents a certain fixed amount of energy. Increasing the frequency would thus increase the total amount of energy in any given strip of highway. When the cars are 1000 feet apart, there would be about five cars per mile of highway. In this analogy, the five cars represent five units of energy. If the distance between the cars is decreased to 500 feet, the frequency is increased, as we just saw. There are now about ten cars per mile of highway, equivalent to ten units of energy. This situation is analogous to that found with different wavelengths of light. As the frequency increases, so does the amount of energy conveyed. A change in wavelength in one direction corresponds to a change in *both* frequency and energy in the opposite direction.

The visible spectrum represents only wavelengths that the human eye can see. Actually, the complete spectrum extends much farther on either end of the visible wavelengths. Figure 1–6 shows the entire spectrum of radiant energy. Beyond the red end of the visible spectrum are encountered longer *infrared waves.* Beyond these are the still longer *microwaves.* Finally, there are *radio waves,* the longest known. Some radio waves have wavelengths of several miles.

Below the violet region of the visible spectrum lie waves of increasingly shorter wavelengths. Immediately beyond violet are *ultraviolet* waves. The small number of ultraviolet waves that penetrate the earth's atmosphere cause sunburn. Beyond the ultraviolet is a range of still shorter wavelengths known as *x-rays.* These waves convey large amounts of energy. They can be quite damaging if they come into contact with living tissue for an extended period of time. Among the shortest wavelengths known are those of gamma and cosmic rays. In gamma rays, the distance from the crest of one wave to the crest of the next may be 0.00652 millimicrons, or about one billionth of an inch. Such a short wavelength means an extremely high frequency. Thus these are extremely high energy radiations.

The wave theory is quite useful in dealing with questions of how light travels, how it is bent (*reflected* or *refracted*), and how one color differs from another. But light presents further characteristics which the wave theory does not explain. For example, when light interacts with matter such as the chlorophyll of green plants, what occurs is better explained by thinking of light as traveling in tiny packets of energy, rather than as waves. These tiny packets of energy are *quanta,* or *photons* (see beginning of Section 1–5). Here, then, is a place where the quantum theory works better than the wave theory. The quantum theory thus provides a second way of describing and understanding the behavior of light.

Quanta, or photons, are thought to be particles which have no mass. Hence, they are not considered to be matter. A photon is thus a "particle of energy." If this sounds contradictory, keep in mind that neither the wave theory nor the quantum theory attempts to explain the true physical nature of light. Both concepts are simply *models* which enable scientists to predict the *behavior* of light under different sets of circumstances. Neither theory claims to explain all of the characteristics of light or to explain what light is. This feature of predicting without explaining is an important aspect of many scientific theories. Most scientific workers would prefer a theory or hypothesis which both predicted *and* gave a consistent picture of reality. Lacking the latter, the former is at least a step away from total ignorance. The degree to which a theory represents the real world is often far less important than the convenient means it gives a scientist for organizing his thoughts and experiments.

Under carefully controlled conditions, and within certain limits, the amount of sugar that a green plant produces is directly proportional to the amount of light energy supplied to the leaves. It has been shown that the green substance chlorophyll absorbs the light energy and converts it into a form of chemical energy. This chemical energy can then be used to produce sugars. The photon concept is useful because it helps us understand how green plant matter can absorb light and use its energy to perform work. For *it is to this ability that man and almost all living organisms owe their very existence.*

According to the quantum theory, the way in which green plants use light energy is as follows: The chlorophyll absorbs photons of light. These photons react with chlorophyll, transfering their energy in definite little "packages." The chlorophyll then converts this energy into a form which can be used by the plant. Within certain limits, the more photons which are delivered to a leaf, the more energy the chlorophyll can absorb. Thus the more sugar that can be produced. By knowing the chemical characteristics of chlorophyll, and the wavelength of light to which it is exposed, it is often possible to predict how much sugar will be produced within given periods of time.

The exact interaction between energy and matter which allows food manufacture in green plants is only just beginning to be understood. It is easy to draw a line of distinction between matter and energy only at certain levels of investigation. The conversion of matter into energy, mentioned at the very start of this chapter, shows that such a basic distinction is mostly a matter of convenience. By taking advantage of this convenience, however, a working foundation of useful concepts is provided on which the exciting superstructure of modern biology can be built.

Chapter 2

The Structure of Matter

2–1 INTRODUCTION

From the time of ancient Greece, through most of the nineteenth century, men debated a seemingly unanswerable question: "What is the fundamental way in which matter is constructed?"

Opinions generally fell into two groups. One group felt that matter could be divided indefinitely into increasingly smaller parts without changing its essential characteristics. The other group thought that there was a limit beyond which matter could not be further divided without changing its properties. The ultimate units to which matter could be reduced were called *atoms*. These atoms were described as being like invisible grains of sand: hard, discrete, and impenetrable.

The question was not easily answered. However, the concept of atoms explained much of the evidence about chemical reactions collected during the seventeenth and eighteenth centuries. Thus, more and more scientists adopted the atomic hypothesis as a working model. Briefly, this hypothesis proposes that all matter is composed of particles called atoms. Furthermore, the atoms of any element, such as oxygen, gold, or aluminum, are different from the atoms of all others. It is this difference in atoms which gives each element its unique characteristics.

2–2 ELEMENTS, COMPOUNDS, AND MIXTURES

Before discussing the structure of the atom, it will be helpful to consider some overall characteristics of matter. Matter can be classified in several ways, as Fig. 2–1 indicates. All matter can be classified as either homogeneous or heterogeneous. To say that matter is homogeneous is to say that a sample of that matter taken from any location, at any time, will be exactly like every other sample. All such samples will have the same characteristics when examined in the laboratory.

To say that matter is heterogeneous is to say that identical samples cannot be obtained from each and every location. Heterogeneous matter may change in composition with time; therefore samples taken at different times will display different characteristics. Matter most frequently encountered in nature is heterogeneous, although there are exceptions.

An example of homogeneous matter is a bar of 100% pure iron. Samples taken from different places at different times will show identical properties. Unrefined iron ore, on the other hand, is an example of heterogeneous matter. A sample from one surface of a piece of ore may show an iron content of 30%. A sample from a different site on the same piece may show an iron content of 60%. It is clear that the samples are not identical. Other substances are present, and these and the iron are not uniformly distributed throughout the ore. This will result in the samples having quite different characteristics when examined in the laboratory.

Other examples of homogeneous and heterogeneous matter are not difficult to find. Salt water is an example of the former, while a mixture of salt water and sand is an example of the latter. If we shake a glass containing salt water and take samples from the surface every five minutes for twenty minutes, all the samples will be identical. If we perform the same experiment with the mixture of sand and salt water, quite different results will be obtained. These four samples will have quite different properties. The most obvious difference is in their sand content. The sand, of course, rapidly settles to the bottom after shaking. This is an example of how the properties of heterogeneous matter may change with the passage of time.

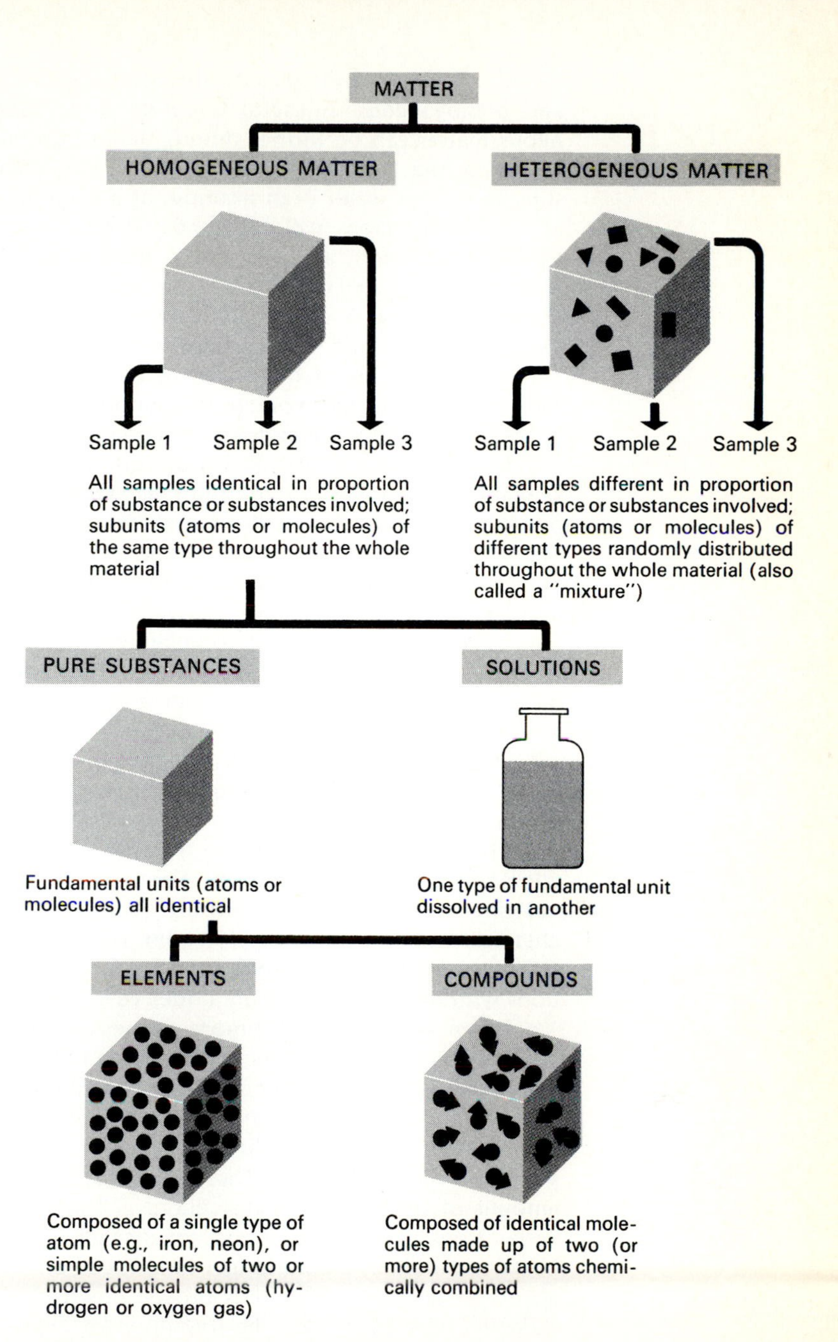

2-1
Diagrammatic representation of the various categories of matter, showing the distinctions based on composition and distribution of fundamental units (atoms or molecules).

The division of matter into these two categories (heterogeneous and homogeneous) is useful but not entirely adequate. Homogeneous matter can be further differentiated into *pure substances* and *solutions* (see Fig. 2–1). The 100% iron bar is an example of pure substance. Salt water is an example of a solution—a solution of salt in water. (Solutions will be discussed later in Section 2–4.) In a further subdivision, all pure substances can be of two types—elements and compounds. The categorical relationship between these various groupings is diagrammed below in Fig. 2–1. Let us now attempt to understand some of the differences between the various categories in terms of the structure of matter. To do this we will have to accept for the moment the hypothesis that *atoms* exist; some of the support for this hypothesis will be discussed later.

If we accept the atomic hypothesis, we can define an element as matter composed of identical atoms. Gold, silver, oxygen, carbon, hydrogen, nitrogen, and sulfur are all examples of elements. A copper strip, if pure, contains only atoms of copper. A lump of gold, carefully refined, contains only atoms of gold.

Today, 103 elements are known. Some occur naturally; others are artificially produced. All 103 elements possess certain characteristics by which they may be recognized. These characteristics are the *properties* of the elements. Properties are divided into two kinds, *physical properties* and *chemical properties.* Some identifying physical properties are weight, density, and conductivity of heat or electricity. The temperatures at which freezing or boiling occur are also physical properties.

Chemical properties refer to the way in which an element reacts chemically with other elements. Some elements, such as sodium or chlorine, are very active chemically. This means that they will form chemical combinations with certain other elements quite easily. Other elements, such as xenon or krypton, form chemical combinations with other elements only under very special conditions.

When atoms of two or more elements combine chemically, the resulting product is a *compound.* For example, two hydrogen atoms unite with one oxygen atom to form the compound water. The basic unit of a compound is the *molecule.** In the case of water, its molecule consists of three atoms. The atoms in a molecule are held together by *chemical bonds.* These chemical bonds represent a certain amount of potential chemical energy. The amount of this energy varies from one kind of bond to another. If a chemical bond is broken, the atoms are no longer held together, energy is liberated or absorbed, and the molecule is broken down. *The building of molecules thus depends upon the formation of chemical bonds between atoms.*

Molecules need not always be formed between different kinds of atoms. The element oxygen, for example, exists as two atoms joined

---

* In ionic compounds (see p. 39) the basic unit is called an *ion pair.*

to form a *diatomic* molecule. The oxygen atoms are joined by a chemical bond. The chemical symbols O_2, H_2, and N_2 are used for oxygen, hydrogen, and nitrogen respectively to indicate that these elements exist in diatomic form. Other elements exist in molecules of more than two atoms. For example, sulfur is usually found with eight atoms, all bound together into one S_8 molecule.

We can now understand certain heterogeneous mixtures such as sand and water. The grains of sand do not form chemical bonds with the water molecules. It is a characteristic of mixtures that the parts composing them can be separated by ordinary physical means. Water and sand, for example, can be separated by straining through a cloth. In a mixture of iron filings and sulfur powder, a magnet can be used to draw out the iron. These are physical processes rather than chemical processes.

None of the properties of the atoms or molecules in matter are changed by being put into a mixture. The fact that no chemical change occurs in their formation is the chief difference between mixtures and compounds. Mixtures are exceptionally important in biological processes. Many important parts of living cells exist in this state. In fact living matter is the most complex mixture of all.

2–3 THE STATES OF MATTER

Matter exists in three main states, *solid, liquid,* or *gas*. At room temperature, some substances are solids (steel), others are liquids (water and mercury), while still others are gases (hydrogen, oxygen, and nitrogen). The state of matter may be changed by raising or lowering the temperature or by varying the pressure. Heat applied to water causes it to boil and vaporize. Thus it changes from a liquid to a gaseous state. When water loses heat, it passes to the solid state, which we know as ice. These changes do *not* involve the making or breaking of the chemical bonds that hold the molecule together. Water always has the molecular formula H_2O whether it is a gas, a liquid, or a solid. For this reason, a change in the state of matter represents a physical rather than a chemical change.

The changes which matter undergoes in passing from one state to another are the result of changes in the motion of atoms or molecules. In a lump of ice, for example, the molecules are constantly vibrating back and forth. As more heat energy is applied, the rate and amplitude of vibration become greater. This means that each molecule not only begins to move faster but also moves farther in each vibration. Above 0°C their motion is too violent to retain the solid state. The ice melts to liquid. As the water passed from the solid to the liquid state, physical attractive forces between the molecules have been overcome by the heat (kinetic) energy supplied. As the temperature is raised still further, the molecules continue to increase their motion. At 100°C the molecules overcome atmospheric pressure and move off into space as a gas. This is what happens when water boils.

Even below boiling temperature, some of the faster molecules are constantly escaping from the water's surface. This is called evaporation. Since it is the faster molecules which are escaping, the net result of evaporation is a loss of heat energy—i.e., cooling.

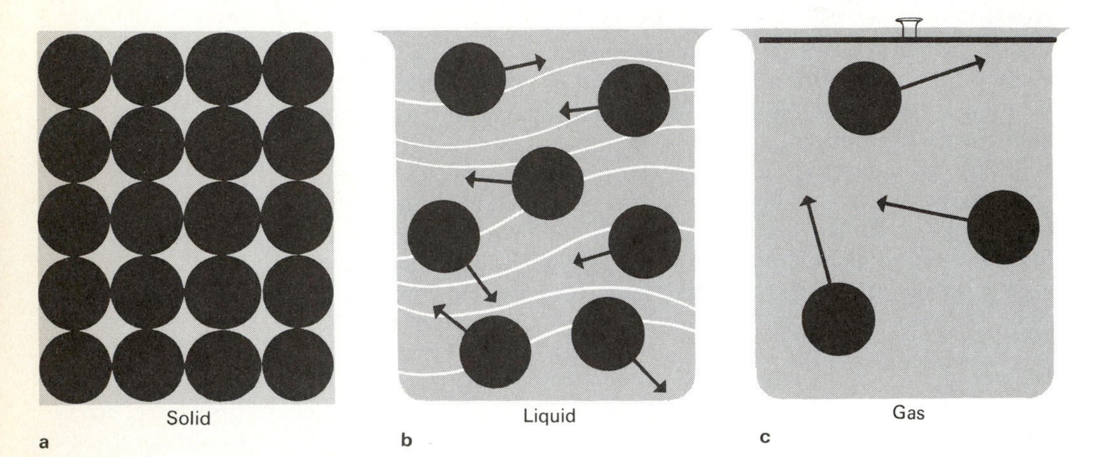

2-2
The density and kinetic energy of atoms or molecules (represented by dots) in a solid, liquid, and gas. Consider that the above diagram represents water in the solid (ice), liquid (water), and gaseous (water vapor) states. In ice the molecules are closely packed and show relatively little motion. They have little kinetic energy. In liquid water the molecules are farther apart and show considerably more motion. The molecules have more kinetic energy than when in the solid state. As water vapor, the molecules are much farther apart and are moving very rapidly. Molecules in the gaseous state have more kinetic energy than those in the liquid or solid states.

2–4 DISPERSION SYSTEMS

When particles of one kind of matter are spread through the particles of another, a dispersion system is created. Such dispersions may be either homogeneous or heterogeneous. The spreading of dust particles throughout the air in a room is an example of a dispersion system. A solid (dust) is dispersed throughout a mixture of gases (air). Other examples of dispersion systems include carbon dioxide gas dissolved in water to form carbonated beverages (a gas dispersed throughout a liquid), gasoline vapor in the air (a gas dispersed throughout a gas), or mist (particles of liquid water dispersed throughout a gas).

Those dispersion systems most important to living systems can be described in terms of various substances dispersed throughout a liquid. The liquid is usually water. The chemical and physical properties of such dispersion systems are determined by both the nature and the size of the particles spread through the water. The formation of a dispersion system does not involve the chemical interaction of the various particles. Therefore, the various substances can be separated from one another by physical means.

Consider the example of salt crystals dissolved in water. When dissolving, particles of the salt spread out between particles of water. The large crystals break down into tiny particles, which cannot be seen. Under ordinary conditions the salt crystals will never settle out of the liquid medium, no matter how long the container is left undisturbed. Such a dispersion system is known as a *solution*. A solution can be defined as a homogeneous molecular mixture of two or more different substances. One of these substances is the dissolving medium. In biological systems this liquid is water. All the substances involved are evenly distributed with respect to one another. However, the proportions can vary in a number of ways. For example, a salt solution can be composed of one gram of salt in fifty, seventy-five, one hundred, two hundred, or one thousand milliliters of water.

The term *solute* is used to refer to the substance being dissolved. The dissolving medium is known as the *solvent*. In the example above, salt is the solute and water the solvent.

Egg white is an example of another type of solution. It is composed of the protein albumen dispersed throughout water. The protein molecule is very large and has certain physical and chemical properties which are different from many types of small molecules. For this reason, the solution takes on a particularly viscous quality. Thus, different types of solutions can be distinguished by the sizes of the particles dispersed throughout the solvent.

In some cases particles in a liquid medium are so large that they will settle to the bottom of the container due to the force of gravity. Sand stirred in water is an example of such a system. Since the particles of sand do not remain evenly distributed among particles of water, this type of system is known as a *suspension*. The difference between a solution and a suspension is partly due to the size of particles in the medium.

Various physical means can be used to separate two or more substances forming a solution or suspension. Salt can be recovered from water by evaporation. By spinning solutions or suspensions at high speeds in a *centrifuge*, particles are forced to settle to the bottom. A centrifuge, in effect, increases the force of gravity on particles in solution or suspension. The faster the centrifuge revolves, the greater the force on the particles. In general, particles are forced to settle out at a given speed of revolution according to their mass. It is thus possible to use the centrifuge to separate different particles mixed together in the same medium. The centrifuge is particularly useful in separating large molecules from each other.

Living matter contains water in which are dissolved or suspended salts, proteins, fats, and other substances. A number of larger cell components such as mitochondria, chloroplasts, and ribosomes are suspended within the liquid portion of cells. In living cells, matter is in constant motion. Thus these larger structures do not settle out of the medium.

2–5 THE ATOMIC NUCLEUS

Atoms are composed of three primary building blocks, *protons, neutrons,* and *electrons.* The only exception is hydrogen-1 (protium), the lightest element, which has no neutrons.

Early in this century, it was suggested that atoms were composed of a small, dense, central portion, the *nucleus,* which in turn was surrounded by various numbers of other particles, the *electrons.* In 1913, Niels Bohr (1885–1962) suggested that the atom resembled a tiny solar system, with the nucleus representing the sun and the electrons the orbiting planets. Later, the nucleus was shown to contain two types of particles, *protons,* which are positively charged, and *neutrons,* which bear no charge. Thus the nucleus was seen to bear a positive charge, due to the protons. The electrons circling the nucleus were found to have a negative charge, offsetting the positive charge of the protons (Fig. 2–3b).

The modern picture of the atom is quite different from Bohr's original picture. The electrons are now considered to form a "cloud" around the nucleus. Figure 2–3a shows this general conception of the atom.

Compared with the electrons, protons and neutrons are very heavy particles. They have almost 2000 times the mass of an electron. Thus, in any atom, the protons and neutrons contribute nearly all the total mass of the atom. When you step on the scales to weigh yourself, only an ounce or two of the weight recorded is due to the electrons of your atoms.

TABLE 2–1

Particle	Mass, gm	Electrical charge	Position in atom
Proton	1.673×10^{-24}	+	Nucleus
Neutron	1.675×10^{-24}	0	Nucleus
Electron	9×10^{-28}	−	Surrounding nucleus

The atomic nucleus, then, is composed of protons and neutrons held tightly together. The energy holding them together is called *binding energy*. The precise nature of nuclear binding energy is only partially understood. Nor has the exact arrangement of the protons and neutrons within the nucleus been definitely determined. However, binding energy plays little role in the chemistry of living organisms. Therefore, it is only of passing interest here.

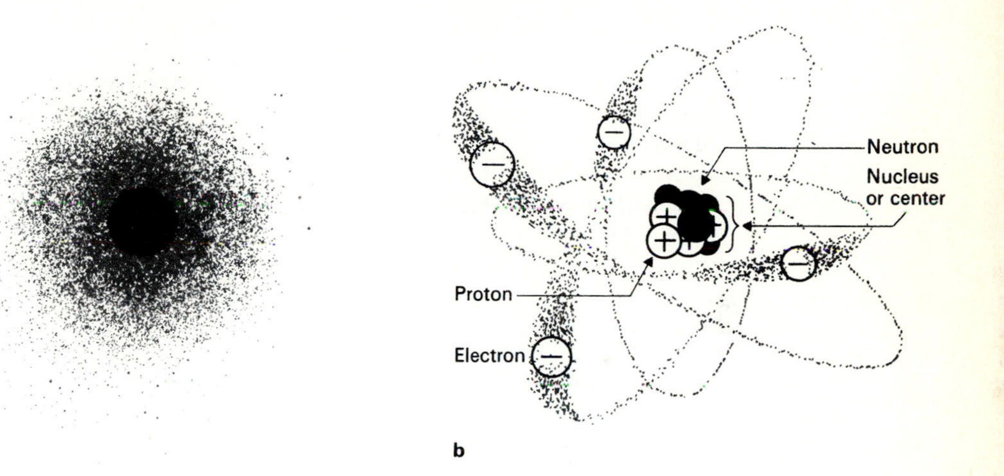

2-3
a) The modern conception of the atom, showing the dense, centrally located nucleus and the outer haze of electrons. (b) Diagrammatic sketch of the atom showing its parts. Negatively charged electrons circle the nucleus, composed of protons and neutrons. This diagram represents a working *model* of the atom. It does not in any real sense represent a "picture" of the atom.

2-6 THE ELECTRONS

Electrons are found as a negatively charged cloud of particles outside of the positively charged nucleus of an atom. They move about the nucleus at varying distances from it, traveling at a high velocity. *It is the electrons which are most directly involved in chemical reactions.* The electron distribution determines which atoms will combine with each other and how many will combine. Therefore, they require a more thorough treatment here than the protons or neutrons of the atomic nucleus.

Electrically neutral atoms have equal numbers of electrons and protons. Under certain conditions, however, an atom can gain or lose electrons. In this way, it acquires a negative or positive charge. When atoms interact with each other, by giving up, taking on, or even sharing electrons, a *chemical reaction* has occurred. The exchanges and interactions of electrons among atoms is the very basis of chemical reactions, and thus of all life processes.

What is an electron? Trying to define an electron is rather like trying to define a sports car. To an avid automobile enthusiast, a written description of a showroom model is not enough. He wants to get the "feel" of the car; how fast it will go; how it will take the turns; its acceleration, and so on. To the scientist, the same is true of the electron. While it can be described in terms of its mass and electric charge, this is only a small part of the total picture.

Important considerations in defining an electron involve knowing its distance from the nucleus, its relationship with other electrons, its characteristic path around the nucleus, and its individual spin on its own axis. An electron can thus be defined only in terms of the way that it acts under certain conditions. Enough is now known about the behavior of electrons to state that no simple definition is possible.

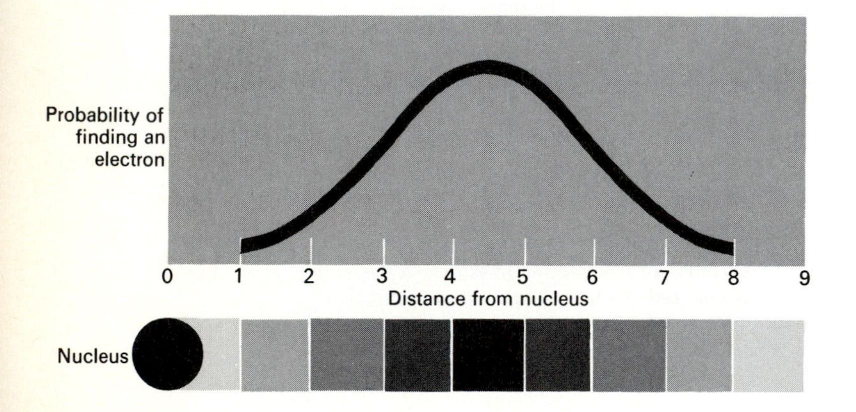

2–4
A graph showing the distribution of electrons in an atom, based on the probability of finding an electron compared to the distance from the nucleus. Below the graph is a physical model which shows electron paths compared with the distances shown on the graph.

Electrons are known to move around the atomic nucleus. However, the practical problem of measuring such movement is so great that descriptions are based upon statistical information. In other words, electron position is spoken of only in terms of probability. It is impossible to predict where a given electron can be located at any moment around an atomic nucleus. It is only possible to predict where certain numbers of electrons will *probably* be found.

In order to describe the pathway of an electron around the nucleus, it is necessary to know its location and velocity at any given moment. By experimentation, physicists can determine either the position of an electron or the velocity. However, they have not been able to determine *both* the position *and* the velocity *at the same time*. Because of the small mass of the electron, the very process of making the measurement changes its position or velocity. If a measurement is made to determine an electron's position, its velocity is changed. If a measurement is made to determine the velocity of an electron, its position is changed.

The inability to predict the precise positions and velocities of specific electrons gave rise to the *Heisenberg Uncertainty Principle*. Although referring specifically to the problem of measuring electron position, this principle has fundamental significance to all science. The Uncertainty Principle implies that by the very process of experimental measurement, man changes the conditions under which the experiment is done. As a result, any predictions about the way in which events occur must be based on probabilities, rather than on certainties.

Consider the following example. In attempting to measure the distance between the earth and the moon, scientists have bounced radar signals off the moon's surface. The time it takes for the radar signals to go to the moon and bounce back gives a measure of the distance they have traveled. However, radar signals are a form of energy. They can move matter. When they strike the moon, its position is changed by a very, very slight degree. The process of measurement has resulted in a change in the original conditions of observation. Electrons are far less massive than the moon. When energy is used to detect the position of electrons, therefore, a very important factor is added to the original conditions.

The problem of measuring electron positions emphasizes two important points. First, it confirms the idea that an electron has a very small mass. The fact that the slightest amount of energy disrupts electrons is a good indication that they possess far less mass than do neutrons or protons. Second, it shows how easy it is to move an electron from one position to another around a nucleus. *Electrons can absorb energy and change positions within an atom.* More will be said on this very important point in a moment.

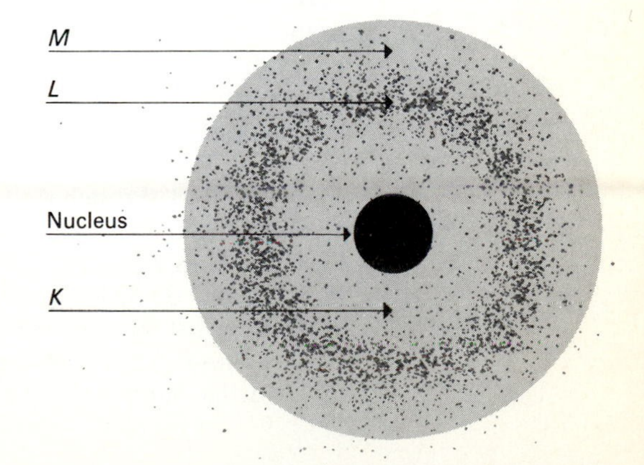

2-5
Diagrammatic representation of an atom, showing three energy levels. *K* represents the lowest energy level, *L* the next higher, and so on. The electrons are shown as a haze since their positions can be determined only in statistical terms.

2–7 ENERGY LEVELS AND ORBITALS

The electron cloud around the nucleus of atoms is composed of a number of different *energy levels*. Energy levels are roughly analogous to the planetary orbits suggested by Niels Bohr in his model of the atom. The term "energy level" is an expressive one, since it leads to thinking of electrons as *particles possessing certain amounts of potential energy*. We might picture electrons outside the atomic nucleus as moving in specific energy levels. The electrons neither absorb nor radiate energy as long as they remain in these energy levels. However, should one or more electrons fall from the energy level which they occupy to a lower one, they will radiate a precise amount of energy. If energy is absorbed by the atom, one or more electrons may jump from a lower energy level to a higher one (see Fig. 2–7).

The amount of energy an electron possesses depends first and foremost upon the energy level it occupies in an atom. The energy levels are analogous to successively higher steps cut into a cliff. The electrons in the energy levels are analogous to rocks of equal size distributed among the various steps, from the ground up. It takes a certain amount of energy to get the individual rocks to each step. Thus the position of each rock represents a certain amount of potential energy.

Orbitals give us a second way of considering electrons in the atom. Again picture the electrons as negatively charged masses (or waves) vibrating in the space around the atomic nucleus. The electrons jump around, following an unknown path. Yet, they are still to be found in a certain region *most of the time*. Imagine a negative charge cloud which describes this high probability region for finding the electron (see Figs. 2–4 and 2–5). These clouds are the orbitals of the atom.

Energy levels and orbitals, then, give us two ways of picturing electron distribution within an atom. The number of electrons to be found in each of an atom's energy levels enables predictions to be made regarding its chemical properties. The orbitals of the atom give us the region where the negative charge density is the greatest, i.e., where the electrons will *most probably* be found.

An atom can have a large number of energy levels. In all atoms it is possible to recognize seven energy levels in which the electrons can be found. These are known as the *K, L, M, N, O, P,* and *Q* levels, or energy levels 1 through 7. Each energy level has a certain maximum number of electrons which it can hold. The seven energy levels are listed in Table 2–2 (page 25) with the maximum number of electrons that actually do exist in isolated, ground-state (i.e., lowest energy state) atoms. More electrons are theoretically possible, the numbers in each energy level being given by the expression $2n^2$, where n is equal to the number of the energy level.

TABLE 2–2

Major energy level	*K*	*L*	*M*	*N*	*O*	*P*	*Q*
Maximum number of electrons	2	8	18	32	32	18	2

If the average distribution of electrons around the nucleus of an atom is plotted on a graph, the result is a curve like that shown in Fig. 2–4. For most atoms, electron distribution increases from the first energy level outward, reaches a peak, and then falls off. This peak is in the second energy level in atoms with small numbers of electrons. When there are large numbers of electrons, as in radium, the peak is not reached until the *N* or *O* level.

An attractive force exists between the negatively charged electrons and the positively charged nucleus of the atom. This force is greatest at the first energy level and falls off in successively more distant levels. This means that electrons in outer energy levels are more easily removed from an atom than those close to the nucleus. In general, it will take less energy to remove an electron from the *Q* energy level than from the *P*, less energy to remove an electron from the *P* than the *O* level, and so forth.

Electrons, then, are attracted to the nucleus of the atom. Hence, like the rocks on the higher steps, electrons farther from the nucleus contain more potential energy.* More energy was required to get them to this greater distance from the nucleus. An electron in an outer energy level releases more energy in falling to the lowest inner level than an electron in an inner level does in falling to the same position. Similarly, a satellite 1000 miles from the earth's surface releases more energy falling to the ground than one which is 500 miles up.

Carrying this last analogy further, it can also be seen that the satellite which is 1000 miles high could be more easily influenced by a gravitational field outside that of the earth, than the one only 500 miles high. In an analogous way, electrons close to the nucleus contain less potential energy but are also less easily attracted away from the atom.

If an atom is stripped of its electrons and is then brought back into contact with an electron source, it regains the lost electrons. In doing so, the innermost energy levels are filled first. The total number of electrons an atom regains under such circumstances depends upon the number of protons in the nucleus since the number of protons equals the number of electrons in an electrically neutral atom.

* This should not be interpreted to mean that the nucleus represents ground zero energy for an electron. Rather, the first orbital represents the lowest potential energy for an electron.

TABLE 2–3

Atomic Number	Element	Energy levels						
		K	*L*	*M*	*N*	*O*	*P*	*Q*
1	Hydrogen	1						
2	Helium	2						
3	Lithium	2	1					
4	Beryllium	2	2					
5	Boron	2	3					
6	Carbon	2	4					
7	Nitrogen	2	5					
8	Oxygen	2	6					
9	Fluorine	2	7					
10	Neon	2	8					
11	Sodium	2	8	1				
12	Magnesium	2	8	2				
13	Aluminum	2	8	3				
14	Silicon	2	8	4				
15	Phosphorus	2	8	5				
16	Sulfur	2	8	6				
17	Chlorine	2	8	7				
18	Argon	2	8	8				
19	Potassium	2	8	8	1			
20	Calcium	2	8	8	2			
21	Scandium	2	8	9	2			
22	Titanium	2	8	10	2			
23	Vanadium	2	8	11	2			
24	Chromium	2	8	13	1			
25	Manganese	2	8	13	2			
26	Iron	2	8	14	2			
27	Cobalt	2	8	15	2			
28	Nickel	2	8	16	2			
29	Copper	2	8	18	1			
30	Zinc	2	8	18	2			
36	Krypton	2	8	18	8			
47	Silver	2	8	18	18	1		
53	Iodine	2	8	18	18	7		
56	Barium	2	8	18	18	8	2	
79	Gold	2	8	18	32	18	1	
92	Uranium	2	8	18	32	21	9	2

Table 2–3 gives electron distribution figures of a number of the more familiar atoms. Examination of this table indicates that atoms of each element have a characteristic *electron configuration.* Electron configuration is the way in which electrons are arranged within that element. *It is this configuration which gives each element its particular chemical properties.* Chemical reactions take place primarily between electrons in the outer energy levels of separate atoms. Consequently, *the number and arrangement of electrons in these levels is important in determining which atoms will react with other atoms.*

The data in Table 2–3 reveal another important characteristic. Note that whatever energy level happens to be outermost in an atom, it never contains more than 8 electrons. For example, the maximum number of electrons in the *M* level is 18. However, in the series of elements between atomic number 11 and 18, the *M* energy level never contains more than 8 electrons. When another electron is added, as in potassium (atomic number 19), it is added to the *N* level. The outermost energy level will never take on more than 8 electrons. For any atom, 8 electrons in the outer orbit represent a stable electron configuration. The importance of this point will become apparent when chemical bonding is discussed in Chapter 3.

2–8 SUPPLYING ENERGY TO ATOMS: ELECTRON TRANSITIONS

Under ordinary conditions, an atom of a given element has all of its energy levels filled in the manner characteristic for that element. It is in an *unexcited* state. In this sense, it is analogous to a pinball machine which is not being played (Fig. 2–6). All the steel balls are in the trough at the lower level of the playing board. Thus they possess the least possible potential energy. If energy is supplied to the balls by a player, they reach a higher energy state. In other words, they gain kinetic energy and are moved out of the tray and go up across the surface of the board.

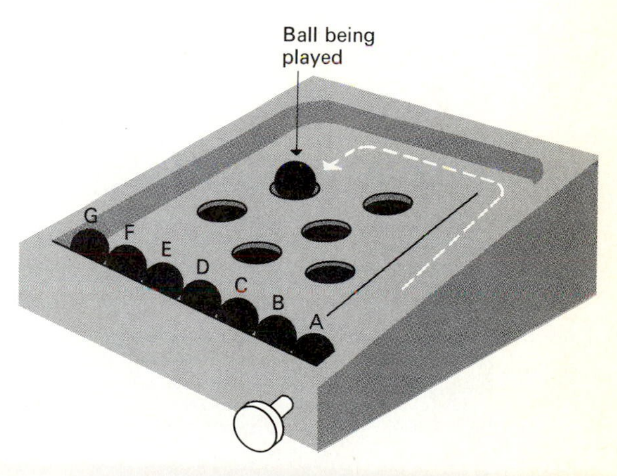

2-6
Diagram illustrating the analogous relationship between a pinball machine and an atom in terms of excitation levels. Pinball at far right is in the lowest energy level for this particular system. When energy is added by the player, the ball can lodge at some location higher on the board. In this location the ball possesses greater potential energy than the balls remaining in the trough. This is analogous to raising electrons to higher energy levels by supplying energy to an atom. The pinball at the highest energy level on top of the board eventually returns to the more stable level in the trough by giving off the same amount of energy as was needed to move it from the lower level.

Eventually, each ball rolls back down across the board's surface and returns to the trough. As the ball rolls, it loses a certain amount of energy. As a matter of fact, it loses exactly the amount needed to raise it that same distance. When all the balls reach the trough again, they are not necessarily in the same order as they were at the start of the game. Some may have taken a longer time to return than others. However, it makes no difference whether or not the order is the same, so long as all the positions in the trough are filled.

This situation is comparable to what happens when energy is supplied to an atom. Increasing the energy of an atom does not just increase its vibrational motion. It also serves to move electrons from lower to higher energy levels. It takes a definite amount of energy to move an electron from one level to another. This process is called *exciting* an atom and can be accomplished by heating to very high temperatures or by electrical discharge. Exciting an atom involves capturing a certain amount of energy for each upward jump of an electron. The term *quantum shift* or *electron transition* is used to describe the movement of electrons from lower to higher energy levels, and vice versa.

In an atom, however, electrons can jump to a large number of energy levels, depending upon the amount of energy supplied. A small, but definite, amount of energy will cause an electron to jump only to the next higher level. The right amount more energy, however, may cause the electron to jump farther, perhaps two or more levels up.

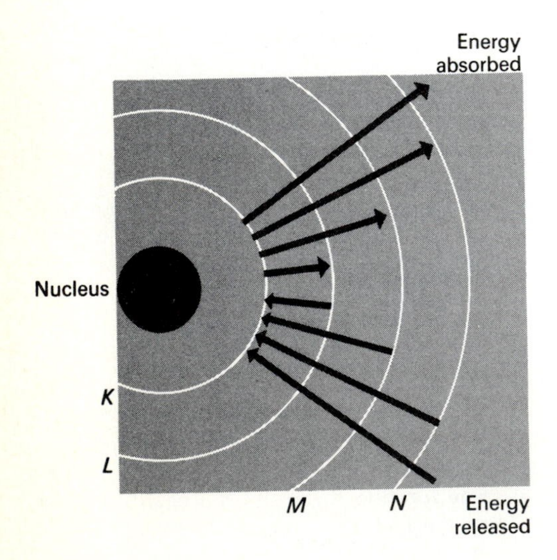

2-7
Electron jumps occur when an atom is supplied with energy. Outward jumps may occur from any energy level to any other, depending on how much energy the electron absorbs. In the above diagram, outward jumps are shown only from the *K* level. When electrons fall inward they give off precise amounts of energy, which appear as x-rays, visible light, and other types of radiant energy. A familiar example of this process is the red neon sign. The neon gas is confined in a closed tube and excited by an electrical discharge. Electrons of the neon atoms absorb some of this energy and undergo specific quantum shifts to higher energy levels. When these electrons fall back to the more stable state, energy is given off in the form of red light.

Raising electrons to a higher energy level produces some gaps below. Some of the lower energy levels are thus incomplete. These are generally filled by other electrons falling down from higher energy levels. *Just as an electron absorbs energy to jump to a higher energy level, so it releases the same amount of energy in falling back to the lower position.*

The movement of electrons to lower energy levels may release energy as x-rays, visible light, or other wavelengths of radiation. The wavelength emitted by an electron in making a downward transition depends upon the distance which it falls and the type of atom in which the transition occurs. To pass from the *M* to the *L* level in atoms of one element involves emitting a certain wavelength of radiant energy. To go from the *L* to the *K* level in the same atom results in a different wavelength. To go from the *M* to the *K* level would produce still a third wavelength. Transitions of the same sort in atoms of other elements would produce a somewhat different series of wavelengths. If light from excited atoms is passed through a spectroscope (a device which bends light through a prism) it is possible to identify many specific electron transitions by the characteristic bright lines which they produce on the spectrum. Every time an electron falls from a higher energy level to a lower one, a specific wavelength of energy is emitted. This shows up as a specific line on the observed spectrum. Such spectrum analysis is based on the principle that for each distance of fall (say from the *M* to the *L* energy level), a specific line will appear on the spectroscope. The number of electrons making any given downward transition in a given period of time (and under specified conditions) determines the brightness of the line. Furthermore, the number and kind of transitions which occur depend upon the electron configuration of the atoms or molecules involved. Hence, study of spectra gives a good clue to the electron structure of atoms, molecules, or ions.

When an electron makes a transition, it absorbs or releases distinct amounts of energy. The quantum theory discussed in Chapter 1 is very useful in dealing with electron transitions, since it describes energy as coming in little packets, or *quanta.* If an atom absorbs one or more quanta of radiant energy, electrons at certain energy levels jump to a higher level. The energy is thus temporarily absorbed by the atom. It may be released in ways which will allow electrons in other atoms to also make transitions. *In this way, energy can be transferred among groups of atoms.*

When atoms unite to form molecules, the energy levels of the individual atoms interact to form *molecular* energy levels. *Electron transitions are possible in molecular energy levels, just as they are among energy levels of individual atoms.* A molecule of chlorophyll, for example, absorbs quanta of light energy from the sun by having electrons raised to higher energy levels. The electrons are temporarily lost to the molecule as a whole. But when the electrons make the downward transitions, they release the same amount of energy they absorbed. This energy is captured by the plant cell, so that it may be used to drive certain chemical reactions which produce simple sugars.

An electron can be caused to jump such a distance from the nucleus that it escapes completely from the atom to which it originally belonged, just as a satellite may escape the earth's gravitational field and go off into space. In the case of the atom, the loss of an electron gives the atom a charge of +1, since it now has one more proton than it has electrons. This electrically charged atom is an *ion*. Loss of two electrons would give the atom a charge of +2, and so on. Ions can also be formed by the gain of electrons. Such ions would be negatively charged. The process of supplying energy to an atom to bring about a loss or gain of electrons is called ionization. This loss or gain of electrons often results in the formation of a more stable electron configuration than the one characteristic of the uncharged atom. For this reason many ions can be regarded as stable chemical substances and do not require a continuous input of energy to sustain the charged state.

2–9 THE IDENTIFICATION OF ATOMS

There are three ways of identifying atoms. The simplest way is to assign each atom an *atomic number. The atomic number of an atom is the number of protons in the nucleus.* Since, in a neutral atom, the number of protons is equal to the number of electrons, the atomic number indirectly tells a great deal about an atom's chemical properties.

Atoms are also identified by *mass numbers.* The proton and neutron are assigned equal masses of one. *The mass number of an atom is the sum of its protons and neutrons.* Because of the small mass of the electron, it is ignored in calculating the mass number of an atom. Atoms of carbon-12, which has 6 protons and 6 neutrons, have an atomic number of 6 and a mass number of 12. To indicate both of these designations, carbon is often written ${}_{6}C^{12}$. An atom of uranium with 92 protons and 146 neutrons has an atomic number of 92 and an atomic mass of 238. This is written ${}_{92}U^{238}$.

Atomic weight is a third identifying feature of an atom. A table of atomic weights and numbers is given in Table 2–4. Such weights are expressed as relative values. This means that they are determined by comparison with the weight of some other atom. Our present system uses carbon-12 as the standard for comparing atomic weights. All other elements are indicated as being so many times heavier or lighter than carbon-12. For example, chlorine is almost three times as heavy as carbon-12, uranium-238 almost 20 times as heavy, and lithium-6 about one-half as heavy.

In determining atomic weight, the use of relative values, rather than the actual mass expressed in grams, is a matter of

TABLE 2–4 Atomic weights based on carbon-12
(Adopted by the International Union of Pure and Applied Chemistry, Jan. 1, 1962)

Element	Symbol	Atomic number	Atomic weight[1]	Element	Symbol	Atomic number	Atomic weight[1]	Element	Symbol	Atomic number	Atomic weight[1]
Actinium	Ac	89	(227)	Gold	Au	79	196.967	Praseodymium	Pr	59	140.907
Aluminum	Al	13	26.9815	Hafnium	Hf	72	178.49	Promethium	Pm	61	147*(145)
Americium	Am	95	241*(243)	Helium	He	2	4.0026	Protactinium	Pa	91	(231
Antimony	Sb	51	121.75	Holmium	Ho	67	164.930	Radium	Ra	88	(226)
Argon	Ar	18	39.948	Hydrogen	H	1	1.00797	Radon	Rn	86	(222)
Arsenic	As	33	74.9216	Indium	In	49	114.82	Rhenium	Re	75	186.2
Astatine	At	85	(210)	Iodine	I	53	126.9044	Rhodium	Rh	45	102.905
Barium	Ba	56	137.34	Iridium	Ir	77	192.2	Rubidium	Rb	37	85.47
Berkelium	Bk	97	249*(247)	Iron	Fe	26	55.847	Ruthenium	Ru	44	101.07
Beryllium	Be	4	9.0122	Krypton	Kr	36	83.80	Samarium	Sm	62	150.35
Bismuth	Bi	83	208.980	Lanthanum	La	57	138.91	Scandium	Sc	21	44.956
Boron	B	5	10.811	Lead	Pb	82	207.19	Selenium	Se	34	78.96
Bromine	Br	35	79.909	Lithium	Li	3	6.939	Silicon	Si	14	28.086
Cadmium	Cd	48	112.40	Lutetium	Lu	71	174.97	Silver	Ag	47	107.870
Calcium	Ca	20	40.08	Magnesium	Mg	12	24.312	Sodium	Na	11	22.9898
Californium	Cf	98	252*(251)	Manganese	Mn	25	54.9380	Strontium	Sr	38	87.62
Carbon	C	6	12.01115	Mendelevium	Md	101		Sulfur	S	16	32.064
Cerium	Ce	58	140.12	Mercury	Hg	80	200.59	Tantalum	Ta	73	180.948
Cesium	Cs	55	132.905	Molybdenum	Mo	42	95.94	Technetium	Tc	43	99*(97)
Chlorine	Cl	17	35.453	Neodymium	Nd	60	144.24	Tellurium	Te	52	127.60
Chromium	Cr	24	51.996	Neon	Ne	10	20.183	Terbium	Tb	65	158.924
Cobalt	Co	27	58.9332	Neptunium	Np	93	(237)	Thallium	Tl	81	204.37
Copper	Cu	29	63.54	Nickel	Ni	28	58.71	Thorium	Th	90	232.038
Curium	Cm	96	242*(247)	Niobium	Nb	41	92.906	Thulium	Tm	69	168.934
Dysprosium	Dy	66	162.50	Nitrogen	N	7	14.0067	Tin	Sn	50	118.69
Einsteinium	Es	99	(254)	Nobelium	No	102		Titanium	Ti	22	47.90
Erbium	Er	68	167.26	Osmium	Os	76	190.2	Tungsten	W	74	183.85
Europium	Eu	63	151.96	Oxygen	O	8	15.9994	Uranium	U	92	238.03
Fermium	Fm	100	(253)	Palladium	Pd	46	106.4	Vanadium	V	23	50.942
Fluorine	F	9	18.9984	Phosphorus	P	15	30.9738	Xenon	Xe	54	131.30
Francium	Fr	87	(223)	Platinum	Pt	78	195.09	Ytterbium	Yb	70	173.04
Gadolinium	Gd	64	157.25	Plutonium	Pu	94	239*(244)	Yttrium	Y	39	88.905
Gallium	Ga	31	69.72	Polonium	Po	84	210*(209)	Zinc	Zn	30	65.37
Germanium	Ge	32	72.59	Potassium	K	19	39.102	Zirconium	Zr	40	91.22

1 For elements available only as radionuclides produced artificially or by decay of longer-lived radionuclides, the mass number of the longest-lived known isotope is listed in parentheses, and the isotope most readily available for chemical experimentation is listed with an asterisk if it differs from the isotope in parentheses.

convenience.* Were the scientist to use the absolute mass of any atom of an element, he would have to manipulate very cumbersome numbers. An oxygen atom, for example, has a mass of 0.0000000000000000000000266 gm.

At first glance it may appear that atomic mass number and atomic weight for any atom are exactly the same. However atomic mass numbers are always whole numbers. Atomic weights, on the other hand, are often given to several decimal places (see Table 2–4). Thus, the atomic mass number for a given carbon atom may be 12 (6 protons + 6 neutrons), but the atomic weight for the *element* carbon is 12.01115. Although for many practical purposes atomic mass numbers and atomic weight values may be treated as though identical, the slight differences between them brings up an important point.

In nature, nearly all known elements exist in several *isotopic forms. Isotopes* are atoms of the same element which have the same atomic number but which differ in atomic mass number. In other words, the atoms have the same number of protons, but vary in number of neutrons. Chlorine, for example, exists in nature as two isotopes. One has an atomic mass number of 35 (17 protons and 18 neutrons), while the other has a mass number of 37 (17 protons and 20 neutrons). The element calcium exists naturally in *eight* isotopic forms, with atomic mass values of 40, 42, 43, 44, 45, 46, 48, and 49. The calcium isotopes thus contain 20 protons plus 20, 22, 23, 24, 25, 26, 28 or 29 neutrons, respectively.

The atomic weight of an element is an *average* of the relative weights for all the naturally occurring isotopes of that element. In averaging these values, the abundance of each isotope in nature is taken into account. With chlorine, 75.4 percent of any sample of this element will be composed of atoms with a relative weight of approximately 35. The other 24.6 percent of any chlorine sample will be composed of atoms with a relative weight of approximately 37. Taking this relative abundance into account, the average atomic weight for a sample of chlorine is thus calculated to be 35.453.†

---

* Though the gram is the basic unit of mass, it is often incorrectly used as a unit of weight as well. In this book, the authors have chosen to use the gram *only* as a unit of mass. Thus, atomic or molecular *weight* is written when units are not given, but atomic or molecular *mass* is specified when grams are expressed. Strictly speaking, an atomic weight is neither a weight nor a mass, but merely the *relative* weight or mass of one atom with respect to a chosen standard (formerly oxygen-16; now carbon-12). Since weight is due to a force (gravity), it would actually be best to use units of force, i.e., *dynes,* to describe it, and to reserve the gram unit for mass only.

† Many isotopes are "unstable." They decompose by radioactive decay to form a more stable atom. Carbon-14 ($_6C^{14}$) has a very slow period of decay: in 5000 years one-half of a sample of carbon-14 has become nitrogen-14. Isotopes of some other elements have much shorter periods of decay, measured in days, hours, minutes, or fractions of a second.

Knowing atomic weights, molecular weights can be found by simply totaling the weights of all the atoms in the molecule. A molecule of sulfuric acid, H_2SO_4, consists of 7 atoms, 2 of hydrogen, 1 of sulfur, and 4 of oxygen. The molecular weight is thus:

$$
\begin{array}{lr}
2 \times \ \ 1 \text{ (atomic weight of H)} = & 2 \\
1 \times 32 \text{ (atomic weight of S)} = & 32 \\
4 \times 16 \text{ (atomic weight of O)} = & 64 \\
\hline
\text{Total:} & 98^{*}
\end{array}
$$

Similarly, the molecular weight for the sugar glucose ($C_6H_{12}O_6$) is the sum of the weights of 6 carbon atoms, 12 hydrogen atoms, and 6 oxygen atoms:

$$
\begin{array}{lr}
6 \times 12 \text{ (atomic weight of C)} = & 72 \\
12 \times \ \ 1 \text{ (atomic weight of H)} = & 12 \\
6 \times 16 \text{ (atomic weight of O)} = & 96 \\
\hline
\text{Total:} & 180
\end{array}
$$

2–10 PHYSICAL AND CHEMICAL CHANGES IN MATTER

The overall types of changes which matter may undergo in living organisms are grouped into *physical* and *chemical changes.*

In a physical change, the form of the matter is changed, but not its chemical properties. The tearing of paper, melting of ice, or dissolving of salt and sugar in water are all examples of physical changes. The results of physical change can often be reversed by ordinary physical means. For example, the ice can be obtained again by simply refreezing the water.

A chemical change involves electron cloud interactions between the atoms of the matter involved. In one such interaction, for example, an atom might give up one outermost electron to another atom. This may result in bonding the two atoms together into a molecule.

The chemical properties of a compound are quite different from those of the elements composing it. Sodium is a very dangerous metal, due to its chemical activity. Chlorine is a poisonous gas. Yet sodium chloride, or table salt, is an essential part of the diet of all higher animals. The chemical characteristics of the sodium and chlorine are changed by the electron interaction which occurred when the

---

* In calculating molecular weights from atomic weights it is customary to round off values to the nearest whole number. Thus, the atomic weight of sulfur (32.064) becomes 32; oxygen (15.994) becomes 16, and so on.

compound was formed. The physical characteristics are also changed, for sodium chloride has none of the physical properties of either sodium or chlorine.

When a chemical change has occurred, the chemical properties of the original reacting substances can only be restored by chemical means. For example, by passing an electric current through some molten sodium chloride, sodium and chlorine atoms can be recovered.

Chemical changes involve far more extensive changes in the nature of matter than do physical changes. Chemical changes involve interactions of electrons. They therefore cause modifications of the chemical characteristics of the elements or compounds involved. Physical changes do not. To the biologist, chemical changes in matter are by and large the most important. All biochemical reactions result in chemical changes. It is thus important to have a general idea of the nature of such changes and the effects they have on the properties of matter.

Chapter 3

The Formation of Molecules

3–1 THE CHEMICAL BOND

Chapter 2 dealt with the characteristics of individual atoms. This chapter will be concerned with particles composed of more than one atom.

Under ordinary conditions, atoms are generally combined with other atoms. Such a combination of two or more atoms is called a *molecule*. The atoms making up a molecule are held together by *chemical bonds*.

When atoms react to form molecules or when molecules break down into atoms, a *chemical reaction* occurs. A chemical reaction involves a *breaking, rearrangement,* and *reforming* of chemical bonds.

In the reaction below, methane and chlorine react to yield the products methyl chloride and hydrogen chloride. A molecule of methane has four hydrogen atoms bonded to a carbon atom (CH_4), and a molecule of chlorine has two chlorine atoms bonded together (Cl_2). When quantities of these molecules are brought together, a chemical reaction occurs. Two bonds are broken: a bond between carbon and hydrogen, and the bond between the chlorine atoms. Two bonds are formed: a bond between the chlorine and carbon atoms, and a bond between hydrogen and chlorine atoms. We can represent this reaction by writing a chemical equation:

$$CH_4 + Cl_2 \longrightarrow CH_3Cl + HCl$$

These bonds can also be represented by a line.

```
   H                           H
   |                           |
H—C—H  +  Cl—Cl  ——→  H—C—Cl  +  H—Cl
   |                           |
   H                           H
```

It is important to realize that all chemical bonds represent a certain stability. It takes an input of energy to break chemical bonds. Bonds are broken only when the opportunity exists to form other more stable bonds. Certain atoms will remain linked together as a molecule indefinitely if other reactive molecules are not around. Molecules may be quite stable until the temperature is increased; then sufficient energy becomes available, making possible the breaking of bonds, and a chemical reaction gets underway. During such a reaction, some bonds will be broken and others formed.

Chemical bonds represent an energy relationship between the atoms making up a molecule. The strength of a chemical bond can be measured in units of energy. The units of energy are frequently expressed as kilocalories (kcal)* per mole of compound.† Table 3–1 gives the bond energies between certain atoms of biological importance. The value given for each bond represents the amount of energy required to *break* that bond. To break one C—H bond, 98.8 kilocalories/mole are required.

$$CH_4 + 98.8 \text{ kilocalories} \longrightarrow CH_3 + H$$

---

* A calorie is the amount of heat energy required to raise the temperature of one gram of water (at 15°C) one degree C. A kilocalorie would thus be the amount of heat required to raise the temperature of 1000 grams of water the same amount. In biology, the kilocalorie is often written as Calorie, with the first letter capitalized to distinguish it from the smaller physical calorie used by physicists and chemists.

† A mole is a measure of the number of particles (i.e., molecules, ions, etc.) of any substance in a given sample of that substance. The number of particles per mole is a multiple of the same constant number for all substances. This constant number, called *Avogadro's number,* is 6.023×10^{23} particles. Thus, one mole of sodium chloride contains $1 \times 6.023 \times 10^{23}$ sodium ions and $1 \times 6.023 \times 10^{23}$ chloride ions. Two moles of sodium chloride contain $2 \times 6.023 \times 10^{23}$ sodium ions and $2 \times 6.023 \times 10^{23}$ chloride ions.

Earlier it was emphasized that in any chemical reaction bonds are both broken and formed. If in the course of any chemical reaction a C—H bond is formed, energy will be given off. The amount of that energy will be 98.8 kilocalories/mole:

$$CH_3 + H \longrightarrow CH_4 + 98.8 \text{ kilocalories}$$

The values from Table 3–1 can be applied to the bonds broken and formed in the reaction discussed earlier: $CH_4 + Cl_2 \rightarrow CH_3Cl + HCl$. Doing this will permit an *overall energy picture* for the reaction to be constructed. From this it will be possible to tell whether energy is consumed or given off during the reaction. In this reaction, two bonds are broken, C—H and Cl—Cl, and two bonds are formed, C—Cl and H—Cl. The energy balance for the reaction can be determined in the following way.

Bonds broken (energy consumed)	
C—H	98.8 kcal/mole
Cl—Cl	58.0 kcal/mole
Total	156.8 kcal/mole

Bonds formed (energy given off)	
C—Cl	78.5 kcal/mole
H—Cl	103.2 kcal/mole
Total	181.7 kcal/mole

The energy given off (181.7 kcal/mole) is greater than the energy consumed (156.8 kcal/mole). Therefore the energy balance lies in the direction of release of energy, and 24.9 kcal/mole are released during this reaction.

The overall picture of the energy balance of chemical reactions is extremely important when considering reactions of biological importance. From such an energy balance we can learn whether energy is required (as in the synthesis of proteins) or obtained (as in the breakdown of ATP).

TABLE 3-1 Some bond energies important in living matter* (Energy required to break the bond indicated).

Bond	Energy kcal/mole
H—H	104.2
C—C	83.1
O—O	33.2
Cl—Cl	58.0
C—H	98.8
N—H	93.4
O—H	110.6
H—Cl	103.2
C—N	69.7
C—O	84.0
C—Cl	78.5

* Data from Linus Pauling, *The Nature of the Chemical Bond,* 3rd edition, Ithaca: Cornell University Press, 1960, p. 85

Remember, although it *always* requires energy to break a bond, there may be an overall release of energy during the reaction if the energy required to break bonds is less than the energy released when new bonds are formed.

The chemical bonds discussed above have been formed as the result of electron interactions among atoms. In the process of a chemical reaction, alterations in these interactions occur and different electron distributions are established for the chemical bonds that are formed. Before a chemical reaction can take place, the reacting molecules must come close enough for their electron clouds to overlap. This means that each atom or molecule must be moving with enough kinetic energy to overcome the natural repulsion of the negative charges of their electron clouds. Atoms which have the proper amount of this kinetic energy can come close enough to allow for this overlapping of electron clouds (see Fig. 3–1).

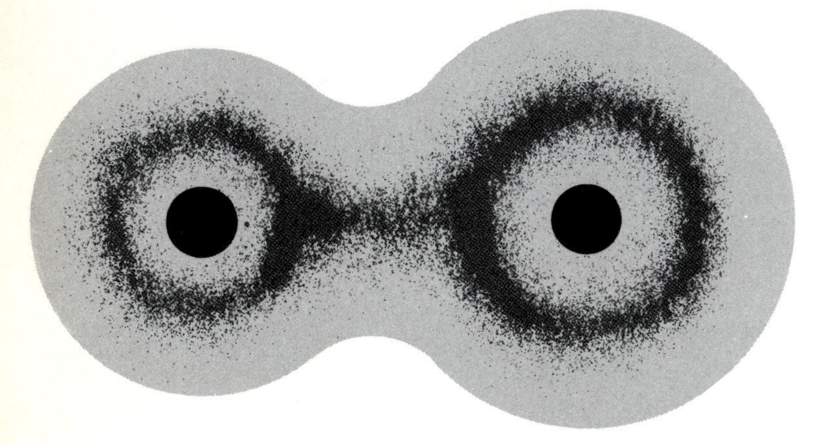

3-1
Interaction of electron clouds of two atoms. This interaction occurs only when the atoms come close enough to each other to allow overlapping of their electron clouds. Chemical reactions result from the rearrangement of electrons between the outer energy levels of the atoms involved.

The overlapping of electron clouds causes rearrangement of electrons in the outermost energy level of each atom. This rearrangement involves one of two possibilities: (1) one atom will *give up* one or more of its electrons to the other, or (2) each atom will *share* one or more electrons with the other. In either case, the total electric charge of one atom may be either less positive or more positive than that of the other. An *electrostatic attraction* is produced between the atoms. This attraction is the chemical bond.

The interaction between outer electrons is the result of a process in which each atom approaches a *stable outer electron configuration*. This phrase requires closer examination. It is central to understanding exactly what we at present think is involved in the formation of a chemical bond.

With the exception of the *K*, or innermost energy level, a stable outer electron configuration is achieved with 8 electrons. The atoms of any element with 8 outer electrons are stable. There is no tendency to accept, give up, or share electrons with other atoms. Such atoms do not react with other atoms. Neon, argon, krypton, xenon, and radon are all examples of such elements. For this reason they were known as *inert* gases. Their lack of chemical activity is thought to be due to their having a stable electron configuration in their energy levels.

The presence of 8 electrons in the outer energy level *forms* a stable condition for most atoms. Isolated atoms in the ground state have no more than 8 electrons in their outer energy levels. Most of those atoms with less than 8 react with other atoms by gaining, losing, or sharing electrons until their outer electron configurations attain stability. The generalization that *chemical reactions between atoms result from the tendency of various atoms to reach a stable outer energy level configuration of electrons* is extremely useful.

There are two major types of chemical bonds found in molecules of biological importance. The distinction between the two is based upon the way in which the stable condition of the outer energy levels is reached.

The first type of chemical bond, found most frequently in inorganic compounds, is the *ionic* or *electrovalent* bond. In the formation of this type of bond, one atom gives up its outermost electrons to one or more other atoms. By doing so, the outermost energy level of each atom becomes more stable. The formation of lithium chloride from the elements lithium and chlorine is an example of ionic bonding (see Fig. 3–2).

The electron configuration for chlorine, counted from the nucleus outward, is 2, 8, and 7. Atoms of lithium have a single electron in the *L* level. In terms of electrons, each lithium atom can reach a stable outer electron configuration of 8 electrons by gaining 7 electrons from each chlorine atom. Or, each lithium atom can donate its *L* electron to a chlorine atom, thereby giving each of the latter a stable outer electron configuration of 8 electrons. The preceding statements, of course, represent thinking carried out only in terms of a hypothetical model; in reality, it is impossible to imagine any separation between the lithium and the thing it is reacting upon.

When lithium gives up its electron, the atom has one less negative charge. Hence it is positively charged (+1). Chlorine, by accepting an electron, now has one more negative charge than positive charges. Thus it is negatively charged (−1). Since opposite charges attract, the positively charged lithium atom and the negatively charged chlorine atom attract each other. This attraction holds the atoms together. A molecule of LiCl is formed.

There is no 100 percent ionic bond. Though one atom tends to give its electrons to another, this "handing over" is not complete. The donated electron may still occasionally circle the nucleus of the donor atom.

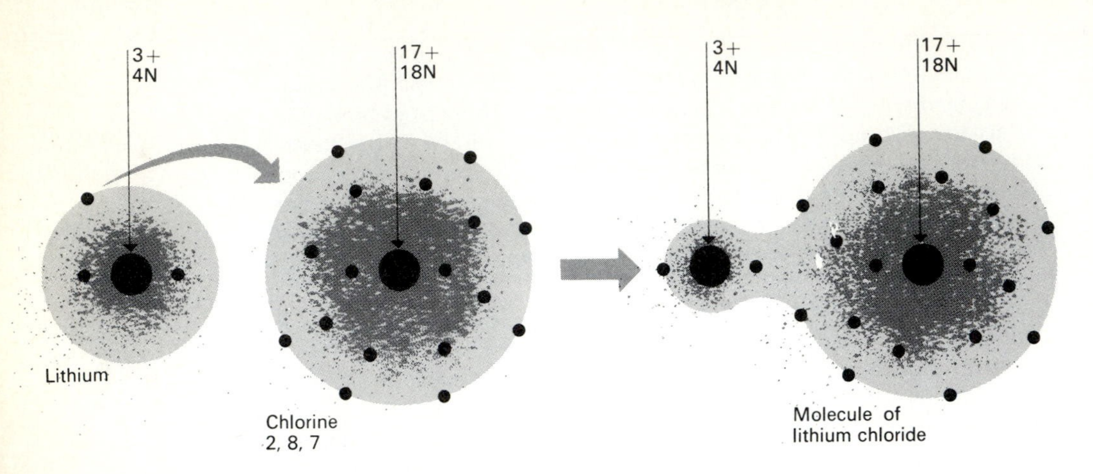

3–2 Formation of an ion pair of lithium chloride.

In the formation of ionic bonds, which atoms will give up electrons and which will receive them? In general, those atoms with fewer than four electrons in the outer energy level tend to lose electrons. Those with more than four tend to gain electrons. Atoms such as sodium, potassium, hydrogen, calcium, and iron possess three or fewer outer electrons. Thus, all of these atoms tend to give up electrons. Atoms such as oxygen, chlorine, sulfur, and iodine need one or two electrons to complete their outer energy levels. Thus, these atoms tend to take on electrons.

What of an atom such as carbon, which has four electrons in its outer energy level? Does such an atom tend to give up or take on electrons when combining with other atoms?

Carbon combines with atoms of many other elements by forming *covalent chemical bonds,* which are a "compromise" between the giving up and the taking on of electrons. *Covalent bonds involve the sharing of one or more pairs of electrons between atoms.* In such bonding, atoms combine by undergoing a rearrangement of electrons in their outer energy levels. Neither atom loses its electrons. Instead, the electrons are shared and may circle the nucleus of any atom in the molecule.

A molecule of the gas methane (Fig. 3–3) illustrates the principle of covalent bonding. Under suitable conditions carbon reacts with hydrogen to form molecules of methane. Four atoms of hydrogen react with each atom of carbon to produce a symmetrical molecule, CH_4:

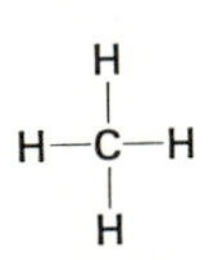

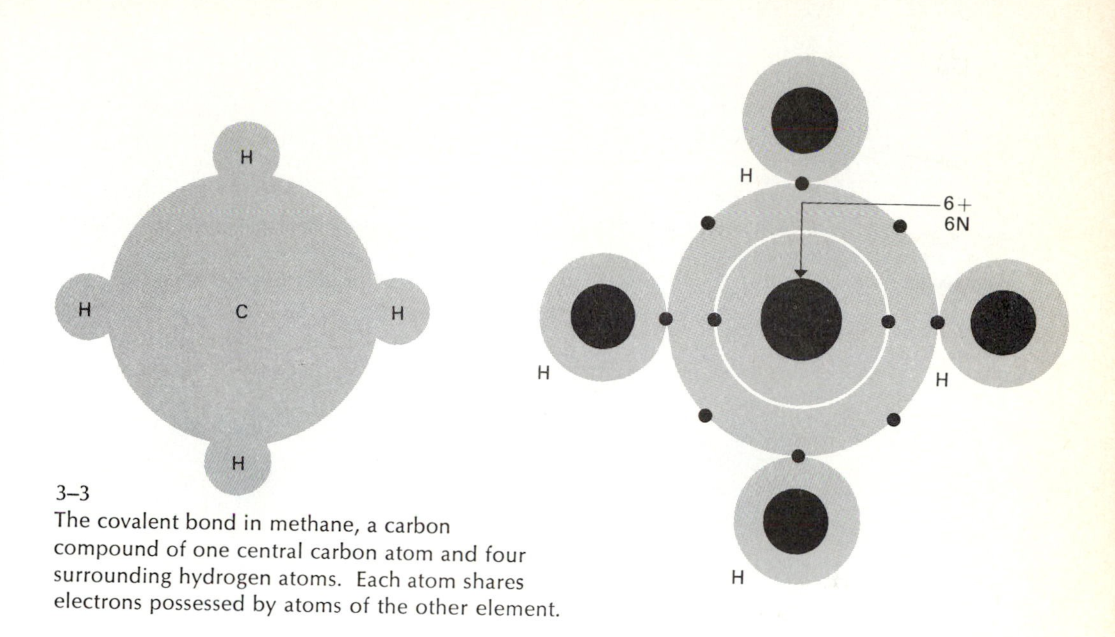

3–3
The covalent bond in methane, a carbon compound of one central carbon atom and four surrounding hydrogen atoms. Each atom shares electrons possessed by atoms of the other element.

Each line between the carbon atom and each hydrogen atom represents a single pair of *shared* electrons. The pair consists of one electron from the carbon, and one from the hydrogen. This may be shown more clearly by writing the molecular formula in the following manner:

$$\begin{array}{c} H \\ \circ\times \\ H \overset{\circ}{\times} C \overset{\circ}{\times} H \\ \circ\times \\ H \end{array}$$

The open dots represent the outer electrons originally in the *L* energy level of carbon. The crosses represent the electrons originally in the *K* level of each hydrogen atom.

Consider why this type of bonding takes place. The carbon atom has four electrons in its outer energy level. To attain stability, carbon needs eight electrons. Each hydrogen atom has one electron in its *K* level. Hydrogen can reach stability by either losing or gaining one electron. In the formation of the covalent bond between carbon and hydrogen, the electrons in the outer energy levels of each atom circle the nuclei of both hydrogen and carbon. As a result, each of the four hydrogen atoms has its own electron plus one electron from the carbon to circle its nucleus. In turn, the carbon atom has not only its own four electrons, but also one from each of the hydrogen atoms to circle its nucleus. This completes the requirements for stability in the outer energy levels of both atoms. It is the sharing of these outer electrons which produces the covalent chemical bond.

3–2 BOND ANGLES AND THE GEOMETRY OF MOLECULES

When three or more atoms combine to form a molecule, there is a definite and predictable geometric relationship established between the atoms involved. In other words, the molecule is given a definite shape. For example, an angle of 104.5° is formed between the two hydrogen atoms and one oxygen atom of a water molecule (Fig. 3–4).

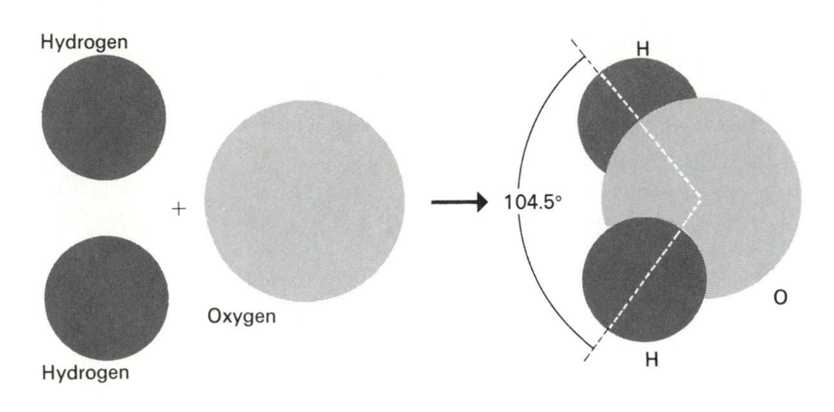

3–4
The formation of water from hydrogen and oxygen. Since the outer electron level of oxygen contains only six electrons and each hydrogen has only one electron to donate, two hydrogen atoms are required to satisfy the stability requirements of the oxygen.

Such angles are called *bond angles*. By certain physical techniques, bond angles can be accurately measured. The determination of bond angles allows the scientist to establish the relative positions of atoms in a molecule.

The way in which one molecule reacts with another depends in part on the shape of each molecule. With small molecules such as those of water, which has a molecular weight of 18, the importance of molecular shape is less apparent. However, with large organic molecules, whose molecular weights range up to several million, shape becomes a crucial factor. The chemical characteristics of some proteins, for example, may be completely changed by simply rearranging one or two atoms.

For the most part, organic compounds are represented in books as if they were flat, two-dimensional structures. In reality, they are three-dimensional. They have depth, in addition to length and breadth. For example, the organic compound methane, discussed previously, is normally written as:

```
    H
    |
H—C—H
    |
    H
```

Actually, the methane molecule forms the outline of a solid, four-sided pyramid known as a *tetrahedron* (Fig. 3–5a, b). There are 109° 28′ bond angles between the four hydrogen atoms, rather than the 90° first indicated on p. 41. It is therefore desirable to show the three-dimensional structure with a "space-filling" model (Fig. 3–5b). Here the space occupied by the atoms, as well as their orientation within the molecule, are taken into consideration.

The concepts of molecular configuration and bond angles play a role in explaining chemical reactions between molecules. These concepts will have importance in later considerations of large molecules found in living organisms.

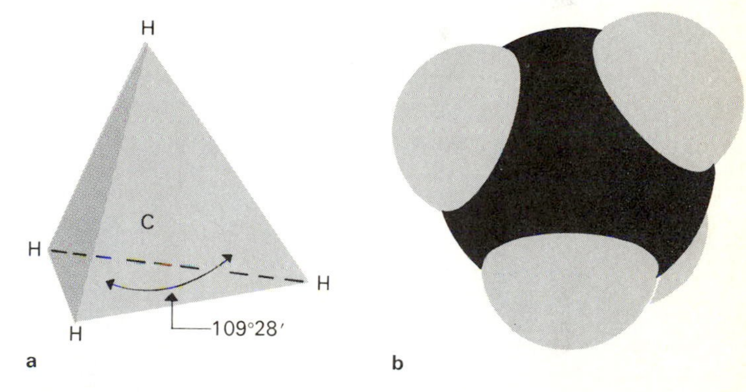

3–5
Two representations of the geometry of a molecule of methane (CH_4). (a) A diagram of the tetrahedronal structure, showing the central carbon surrounded by 4 hydrogens all at equal distances from the carbon and from each other. (b) A space-filling model of methane showing the actual volumes and geometric relations of the atoms. The black atom in the center is carbon. The fourth hydrogen is partly hidden on the other side of the molecule.

3–3 THE POLARITY OF MOLECULES

We have seen that atoms differ in their capacity to "hold onto" their electrons; some gain electrons, some lose electrons. Certain atoms, oxygen and nitrogen, for example, do not have sufficient electron-attractive power to become fully charged negative ions.

However, the attraction for electrons is sufficiently great so that, when covalently bonded to hydrogen, the electrons are not equally shared between the two nuclei. The electrons tend to spend more time around the oxygen nucleus and consequently less time around the hydrogen nucleus. This means that one portion of a molecule is slightly positive or slightly negative in relation to another portion of the same molecule. When such an uneven distribution of charge occurs, the molecule is said to exhibit *polarity.* The molecule has a positive and a negative end, separated from each other like the poles of a bar magnet. Because this is not a full −1 or +1 charge but a smaller charge, it is represented as $\delta+$ or $\delta-$ (delta positive or delta negative).

Water is a good example to illustrate this point. Although, as a whole, the water molecule is electrically neutral, it does have a positive and a negative end (Fig. 3–6). The geometric configuration of the molecule places both hydrogen atoms at one end. The nucleus of the oxygen atom attracts electrons more than the nuclei of the hydrogen atoms. This results in two slightly positively charged regions on one end of the molecule and a single slightly negatively charged region on the other. The molecule thus has a positive and a negative end, or two poles. The water molecule is a *polar molecule.*

The significance of molecular polarity to the biological sciences comes from two main areas: First, polar molecules tend to become oriented with respect to other molecules. Because of this, polar

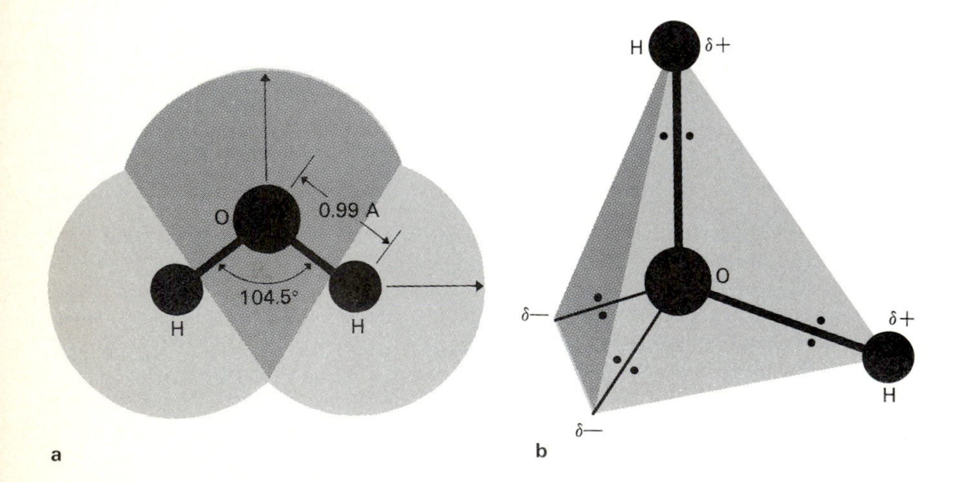

3-6
The structure of water, illustrating its precise geometry and its consequent polar properties. (a) The bond distances and bond angle of the water molecule. (b) The oxygen atom exerts a strong attractive force on the four unpaired outer electrons; as a result there is a separation of positive and negative centers of charge that point exactly at the four corners of the tetrahedron. The consequence is a molecular configuration with a distinct positive and negative region.

molecules are important in helping to establish the three-dimensional structure or orientation of other larger molecules. For example, molecules of fatty acids (Chapter 8), found in all living matter, are composed of a nonpolar carbon chain with a polar carbon-oxygen group (COOH) at one end. When placed in water, the polar ends of the fatty acid molecules are attracted to water molecules, which are also polar. The nonpolar carbon chains are at the same time repelled by the water. As a result, fatty acid molecules are oriented on the water's surface (Fig. 3–7).

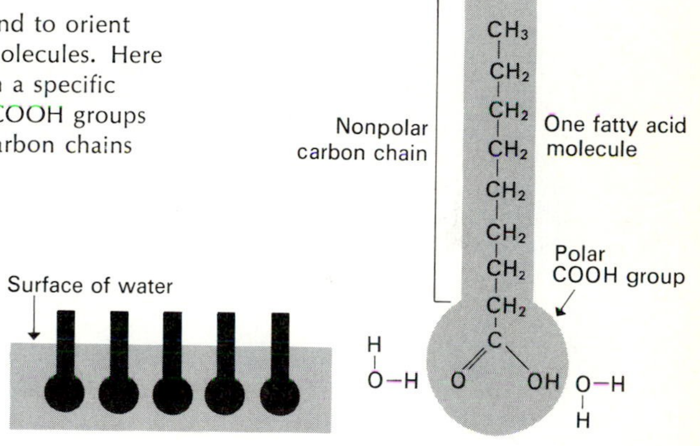

3–7
Polar molecules, such as fatty acids, tend to orient themselves in respect to other polar molecules. Here the molecules of a fatty acid line up in a specific fashion on the surface of water. The COOH groups are in the water (also polar) and the carbon chains stick out into the air.

Of particular importance to living things is the orientation of *phospholipid* molecules, which are a combination of a fat molecule with a phosphate group. Phospholipids are among the most important parts of cell membranes. They tend to become oriented on surface or boundary regions in a manner similar to the fatty acids on water. It is partly in this way that cell membranes are given a distinct structure.

Second, polarity is important in understanding both the geometry and the chemical characteristics of large molecules, such as proteins. Proteins are so large that they may possess a number of *polar groups* on one molecule. Polar groups, like radicals (Section 3–4), are simply groups of atoms which bear as a unit a partial positive or a partial negative charge. The specific geometry of proteins exists in part because polar groups on one part of the molecule attract polar groups on another part of the same molecule. This stabilizes the specific twisting and folding of the molecule which is all-important to the chemical characteristics it displays.

Polarity thus tends to bring small molecules, or specific regions of large molecules, into definite geometric relation. In this way, the chemical bonding between individual molecules or between specific portions of large molecules is brought about more easily.

In living organisms, one of the most common types of chemical bonds produced by polar attraction is the *hydrogen bond*. Hydrogen bonds are produced by the electrostatic attraction between positively charged hydrogen atoms (protons) on one part of the molecule, and negatively charged atoms of oxygen or nitrogen on the same or another molecule. The oxygen and nitrogen atoms are partially negatively charged because their nuclei attract large numbers of electrons around them. Because the hydrogen bond occurs between polar regions of a molecule, it is, like all polar attractions, relatively weak.

A simple example of hydrogen bonding can be seen between water molecules. The hydrogen atoms of one water molecule form a hydrogen bond with the oxygen atom of the adjacent molecule (Fig. 3–8).

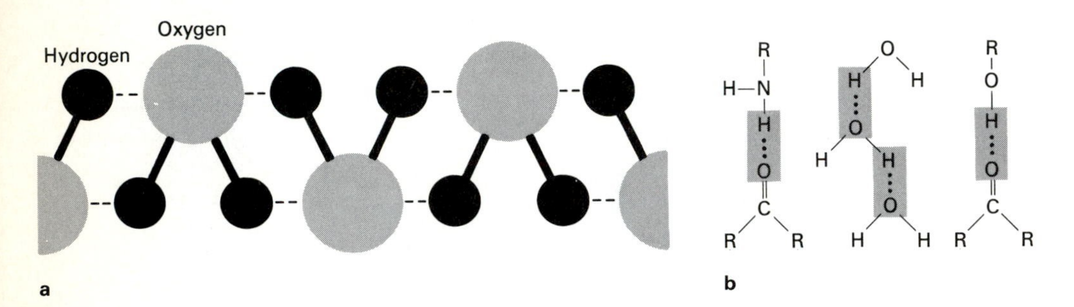

3-8
Water serves as a good example of the polar orientation of molecules which results specifically in hydrogen bonding. (a) Note that molecular shape is an important factor. The hydrogen bonds are represented by -------. (b) Several types of hydrogen bonds are represented by the dotted lines in this diagram. The structure on the left shows bonding between an amino hydrogen and a carbonyl oxygen. The structure on the right shows bonding between a hydroxyl hydrogen and a carbonyl oxygen. The structure in the middle represents hydrogen bonding between molecules of water. Hydrogen bonds (the shaded areas) are not as strong as covalent bonds, but they do contribute a significant degree of stability in the structures. In water, for example, the hydrogen bonds are responsible for water's unusually high boiling point.

3–4
IONS AND COMPLEX IONS (RADICALS)

At room temperature, the compound hydrogen chloride (HCl) is a gas. If molecules of this gas are dissolved in water, the hydrogen is separated from the chlorine. The separation, or *ionization,* occurs in such a way that the hydrogen atom does not take back the electron which it loaned to chlorine in forming the bond. Thus the hydrogen atom, now a proton, bears a charge of +1, and is called a *hydrogen ion*. The chlorine retains the extra electron which it received from the hydrogen atom. Since it has one more electron than protons, the chlorine bears a negative charge of −1. This particle is a *chloride ion*. The hydrogen ion is written as H^+, and the chloride ion as Cl^-.

As we saw in Chapter 2, an ion can be defined as *an atom, or a group of atoms, which bears an electric charge*. This means that ions are formed whenever an atom loses or gains electrons. The dissolving of sodium chloride in water results in a separation, or *dissociation*, of the sodium and chloride ions (see Fig. 3–9). Recall that sodium chloride is an ionic compound; its component atoms are already in the ion state. Molecular substances, on the other hand, such as hydrogen chloride, lithium chloride, acids, etc., undergo ionization when entering into solution. As soon as the water is removed, the oppositely charged particles recombine to form the same molecules or ion pairs.

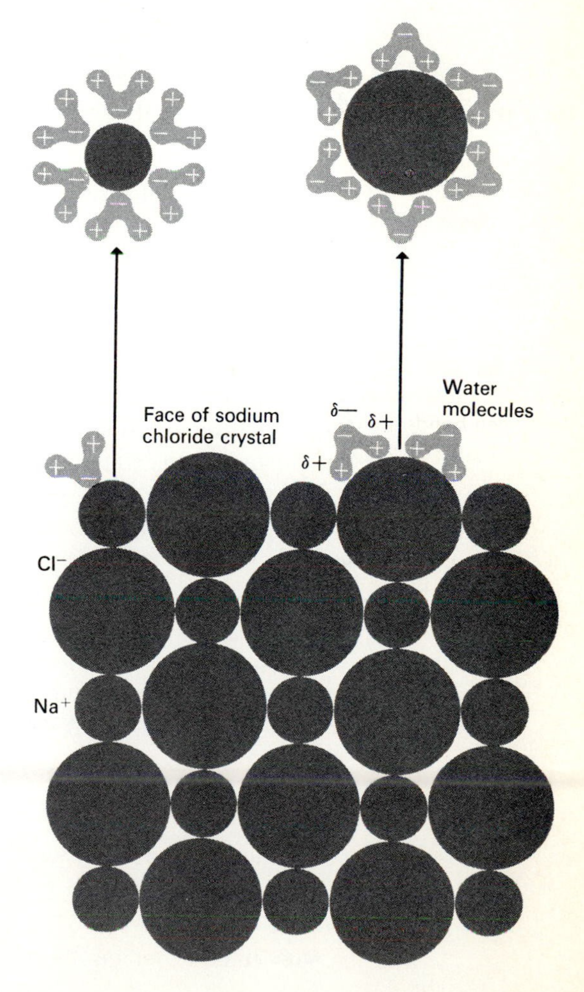

3-9

A representation of the dissolving of sodium chloride in water. The negative surfaces of the polar water molecules are attracted to the positive sodium ions, and pull them off the crystal lattice. The positive surfaces of the water molecules are attracted to the negative chloride ions, and pull them off. As indicated, each sodium and chloride ion is probably surrounded by at least six water molecules.

The process of ionization can be represented by an *ionization equation*. In chemical language, an equation indicates what goes into and what comes out of a certain reaction. For example, the ionization equation for HCl, given below, indicates that one molecule of this compound yields, upon ionization, a positively charged hydrogen ion (H^+) and a negatively charged chloride ion (Cl^-).

$$HCl \rightarrow H^+ + Cl^-$$

Similarly, the dissociation of sodium chloride (table salt) gives

$$NaCl \rightarrow Na^+ + Cl^-$$

When calcium chloride ionizes, it produces two chloride ions for every ion of calcium:

$$CaCl_2 \rightarrow Ca^{2+} + 2Cl^-$$

This equation tells us several things. First, the molecule of calcium chloride consists of one atom of calcium and two atoms of chlorine. The subscript after the symbol for any atom indicates the number of atoms of that element in the molecule. Second, in writing the ionization equation for this compound, we must account for the two atoms of chlorine by showing that there are two chloride ions in the solution. To indicate that these two ions *do* occur, a 2 is placed in front of the Cl^-. The equation is now *balanced*. Each of the atoms in the reaction is accounted for.

When some compounds ionize, one of the products is a complex ion. Complex ions are associations of two or more atoms which bear an overall positive or negative charge. *They are thus collections of two or more atoms which act as a single ion.* For example, the complete ionization of sulfuric acid yields hydrogen ions and a sulfate ion:

$$H_2SO_4 \rightarrow 2H^+ + \underset{\text{(sulfate ion)}}{SO_4^{2-}}$$

The sulfate ion is composed of one atom of sulfur and four atoms of oxygen. These five atoms have an overall electric charge of -2.

Molecules of calcium nitrate contain two nitrate ions bonded to one calcium atom. They are written as:

$$Ca(NO_3)_2$$

Aluminum sulfate, which contains three sulfate ions bonded to two aluminum atoms, is written as:

$$Al_2(SO_4)_3$$

The parentheses enclose the complex ion itself. The subscript number after the parentheses indicates the number of radical groups contained in the molecule.

Not all compounds which ionize in water do so with equal readiness. In all of the ionization equations listed above, nearly 100 percent of the molecules dissociate to release the appropriate ions. However, water molecules are examples of molecules which ionize only very slightly, so that the reaction

$$H_2O \rightarrow \underset{\text{(hydroxide ion)}}{H^+ + OH^-}$$

occurs in approximately one out of every 554 million molecules. Carbonic acid ionizes more than water, but still only about one percent of the molecules dissociate:

$$H_2CO_3 \rightarrow \underset{\text{(bicarbonate ion)}}{H^+ + HCO_3^-}$$

The degree to which a molecule will enter into solution depends upon the type of chemical bond holding the atoms together. In general, the more ionic the bond, the more readily dissolving will occur. Conversely, molecules which are held together primarily by covalent bonds show little tendency to ionize. Sodium chloride and lithium chloride have molecules bound together by an almost totally ionic bond. They readily dissolve in water. Carbon compounds such as methane, however, show almost complete covalent bonding. They show virtually no dissolving in a water solution. Carbon compounds that contain oxygen or nitrogen bonded to hydrogen are often readily soluble in water. This is due to the polar nature of the O—H or N—H bonds, which permits interaction with the polar water molecules.

Atoms which are held together by ionic bonds can separate more easily because one atom has given up electrons while another has accepted them. In this way the outer energy level of each atom has been satisfied. Thus, when such molecules dissociate, there is no further exchange of electrons required. Dissociation in this case merely involves overcoming the electrostatic attraction between positive and negative particles. The action of water molecules accomplishes this dissociation (see Fig. 3–9).

In covalent bonds, the outer energy level of each atom is satisfied only as long as the shared electrons revolve about both nuclei. For this to be possible, the atoms must remain close together. It is very difficult to separate one from another if, in so doing, the atoms are forced to assume unstable outer electron configurations. For this reason water molecules cannot, in general, force apart atoms which are covalently bonded. Such molecules thus fail to show ionization in water solution.

3–5 VALENCE: WRITING MOLECULAR FORMULAS

The net charge on an atom or radical following electron transfer interaction with other atoms or radicals is its *valence*. A positive valence sign means that the atom or molecule has donated electrons. A negative valence sign means that it has received electrons. For example, upon donating an electron to some other atom, a sodium atom becomes a sodium ion with a valence of plus one (+1).

Every element can be described by the change of valence it shows in chemical reactions. These values refer to the number of electrons lost or gained. To say that the valence of hydrogen is +1 means that hydrogen normally undergoes reaction by losing one electron.

Knowing the change in valence which an atom or molecule usually undergoes is useful in predicting the proportions in which elements combine in chemical reactions. For example, the valence of hydrogen is +1, that of chlorine is −1. Thus an electrically neutral molecule can be formed by the interaction of one atom of hydrogen and one of chlorine. The number of electrons which one atom needs to attain a stable outer energy level cannot always be supplied by a single donor atom. However, in the case of hydrogen and chlorine, the match is perfect, since chlorine needs one electron and hydrogen has one to donate. Such combinations as magnesium chloride ($MgCl_2$) or aluminum oxide (Al_2O_3) show that frequently several atoms of one element are needed to satisfy the electron requirements of another.

Two examples will illustrate how it is possible to derive the formula for a compound if we know the usual valence change which particular atoms undergo.

Consider the formation of magnesium chloride from magnesium and chlorine atoms. The valence of magnesium is +2. The valence of chloride is −1. Since most molecules are electrically neutral, the positive and negative values must be equal. It will thus require two chloride ions, each with a charge of −1, to balance the two positive charges on one magnesium atom. The formula for magnesium chloride is thus $MgCl_2$.

The reaction between oxygen and the metal aluminum is a second example. Each atom of aluminum has three outer electrons. The valence of aluminum is thus +3. Oxygen atoms, however, need *two* electrons to become stable. If one aluminum atom combined with one oxygen, there would still be an extra electron. The two-atom unit would bear a negative charge. Therefore, a stable molecule cannot be built from just one atom of aluminum and one of oxygen.

It is evident, however, that 2 atoms of aluminum donate a total of 6 electrons and that 3 atoms of oxygen take on exactly 6 electrons.

The proportions of 2 aluminum atoms to 3 oxygen atoms *could* form an electrically balanced molecule. Experimental analysis of the compound confirms this conclusion. Thus, the molecular formula for aluminum oxide is written as Al_2O_3.*

During chemical reactions atoms combine in definite proportions. If the reaction producing a compound is carried out under precisely the same conditions, the proportions of atoms of one element to those of another within the compound is always the same. The predictable nature of chemical reactions is partly a result of the formation of molecules according to the general principles outlined above.

3–6 VARIABLE VALENCE

Some elements possess two or more different valences. Iron, for example, can exist as the ferrous ion (Fe^{2+}) or as the ferric (Fe^{3+}) ion. Other atoms such as copper, cobalt, and nickel also show this characteristic. In all these elements, one or more electrons can be shifted between an outermost orbital and the next orbital or energy level beneath it. The number of electrons in the outer energy level of these elements, therefore, can vary. Hence, there is a variation in the number of electrons which can be lost or gained during reaction.

In iron, for example, the outer energy level *N* can contain three electrons, while the *M* level contains 13. Or, the *N* level can contain two electrons, while the *M* level has 14. In the first case, the atom would have three electrons to donate. It would thus have a valence of +3. In the second case, the iron would have only two electrons to donate. As an ion, it would have a valence of +2. Whether or not a given atom of iron loses two or three electrons depends upon the atoms with which it is reacting and the conditions under which the reaction takes place.

Both iron and copper compounds function in the transport of oxygen. The red pigment *hemoglobin* has a molecule composed of a ring-shaped structure of carbon atoms, a protein, and a core of four iron atoms. Hemoglobin gives color to the blood of most higher animals. Similar molecules, which contain copper, are found in the blood of certain lower animals where they serve the same function as hemoglobin in higher animals. In addition, iron compounds function as important agents of electron transfer, thus facilitating the release of energy within the cell. The ability of hemoglobin and other iron compounds to carry out such specialized functions is due in part to the ability of iron to shift electrons between two energy levels.

---

* In using valence to determine these molecular formulas, we are assuming that eight electrons is the maximum number for stability.

3–7
OXIDATION AND REDUCTION

The process of losing electrons is called *oxidation.* The atom that loses electrons is said to be oxidized. The process of gaining electrons is called *reduction.* The atom that gains electrons is said to be *reduced.*

The process of oxidation does not necessarily involve the element oxygen. The name "oxidation" was originally derived from the class of reactions involving the combination of various elements (mostly metals) with oxygen. Elements or compounds which combine with oxygen are oxidized. They give up electrons. Now, the term oxidation is used to refer to *any* loss of electrons in a chemical reaction, whether or not oxygen is involved.

Oxidation and reduction are useful terms when employed to describe what happens when two atoms, such as sodium and chlorine, combine to form a compound—in this instance, sodium chloride.

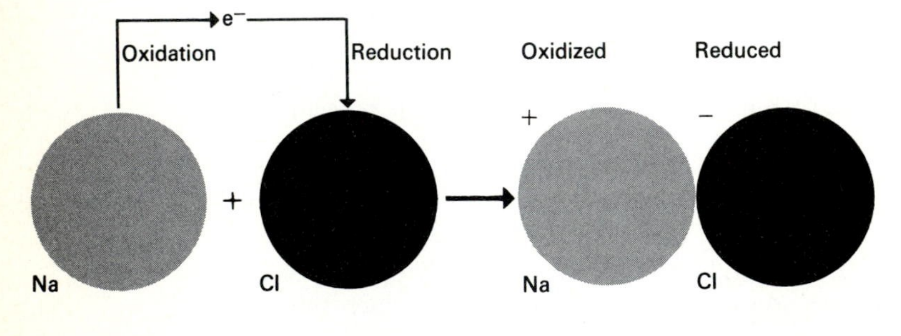

3–10
Electron exchange in oxidation-reduction reactions. In forming sodium chloride, sodium atoms give up one electron to chlorine. The sodium is said to be oxidized; the chlorine, reduced.

Sodium atoms undergo a change from a neutral to an electrically charged condition (from 0 to +1) by losing an electron. Chlorine goes from neutral to −1 by gaining an electron. The sodium is thus oxidized and the chlorine reduced. Thus, the formation of ionic chemical bonds involves an oxidation-reduction reaction. Figure 3–10 shows a diagrammatic representation of the oxidation-reduction relationship in a reaction between sodium and chlorine.

By contributing electrons which reduce chlorine, sodium acts as a *reducing agent.* By accepting electrons from sodium, chlorine acts as an *oxidizing agent.* An oxidizing agent, then, is one that accepts electrons and is itself reduced, while a reducing agent is one that gives up electrons and is itself oxidized.

The concepts of oxidation and reduction are convenient ways of viewing chemical reactions. This is especially true of reactions in biological systems. Here, *the removal of electrons from one substance and their subsequent transfer to others is the chief means of energy release.*

3–8 HIGH-ENERGY BONDS: ATP

When fuel molecules are broken down within living organisms, the energy released must be channeled in a useful direction. If it is not captured immediately, the energy is lost for useful work within the cell.

None of the energy released is used directly to power chemical reactions. All such energy is stored in small "packages" of energy known as *high-energy phosphate bonds.* In this way the energy is available in a common form for all the metabolic processes of the cell.

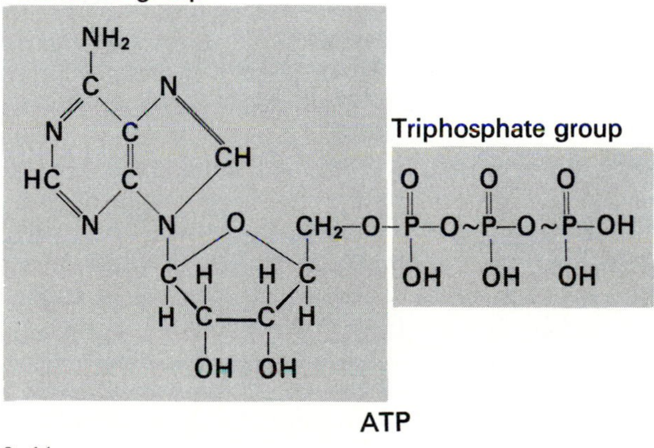

3–11
Structural formula for ATP.

In most living systems, high-energy bonds are found in the form of a compound known as *adenosine triphosphate* (ATP). It is composed of a central molecular unit, adenosine, to which are attached three phosphate groups (Fig. 3–11). ATP is frequently called a "high-energy" compound because a great deal of energy which can be used by the cell is released when one or both of the terminal phosphate groups are removed. Under cellular conditions, about 7.3 kilocalories of energy are released per mole of ATP reacting.* This energy is capable of driving other reactions that require an energy input, for example, protein synthesis. The chemical reaction that results in the removal of a phosphate group is called *hydrolysis* (*hydro* = water; *lysis* = decomposition) because water is a necessary reactant along with ATP; this chemical reaction is shown below with a structural equation:

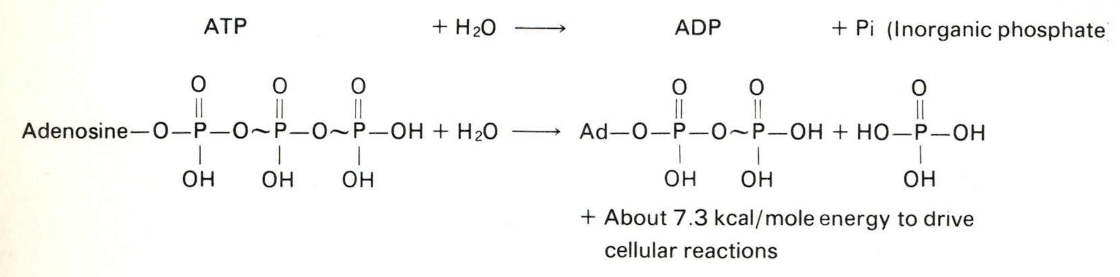

+ About 7.3 kcal/mole energy to drive cellular reactions

The terms *high-energy bond* and *high-energy compound* really refer to the energy released in such a hydrolysis reaction. These terms are somewhat misleading and should not be thought of as indicating either (1) that such bonds necessarily contain more energy than many other bonds (*high-energy bond* does not mean the same thing as the term *bond energy* discussed earlier), or (2) that the energy resides solely in the bonds designated ~ (as between the oxygen and the phosphorus). The "high energy" component of such bonds resides in the complex interaction of a number of atoms in the molecule, and not in the interaction between any two; the use of the ~ between two atoms is only a convenience employed in writing the structural formula

For living organisms there are two advantages to having energy stored in high-energy phosphate bonds. First, the energy in such bonds is readily available to the cell for immediate use. The process of extracting the energy from the phosphate bond is a one-step reaction. Second, and perhaps most important, the amount of energy in a high-energy phosphate bond is approximately that amount which is the most useful for producing biochemical reactions. This means that there is less wastage of energy. As a result, biochemical reactions within living organisms are quite efficient. They do not release more energy than can be used at any one time.

---

* The reader may encounter different values for high-energy phosphate bonds in the literature. The reason for the disparity, which may run from 7.3 kilocalories per mole to 10, rests largely with the conditions under which the measurements in any specific case were made. pH greatly alters the amount of energy released in breaking a high-energy bond from ATP, as does the presence or absence of certain ions, such as magnesium. The standard value which we shall use is 7.3 kilocalories per mole; this measures the energy released by hydrolyzing the terminal phosphate at pH 7.0 (neutrality) with no magnesium present.

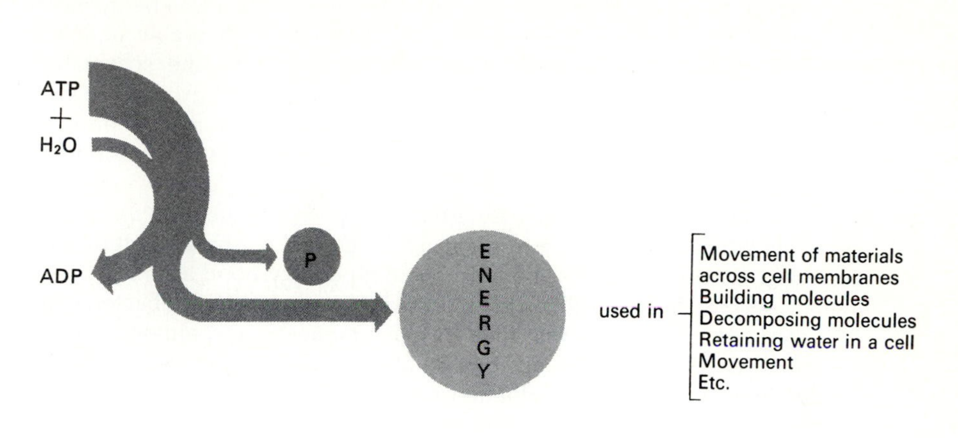

3–12
Metabolic reactions run on the energy derived from converting ATP into ADP or AMP.

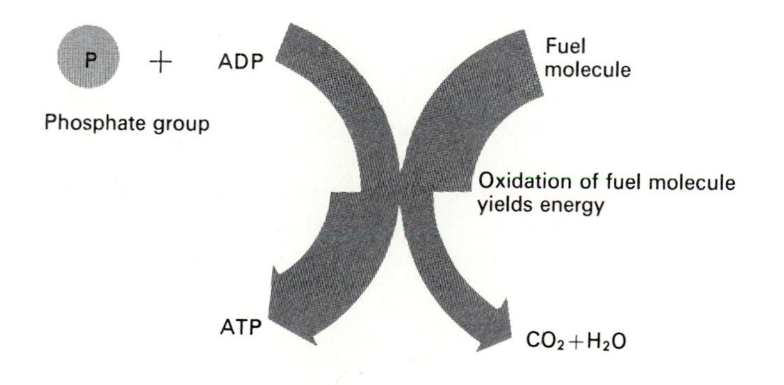

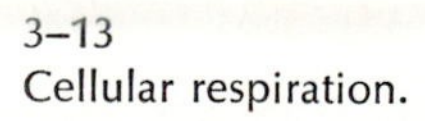

3–13
Cellular respiration.

A high-energy bond results from the internal rearrangement of electrons in a molecule. The electron configuration of the phosphate group is apparently well suited for this rearrangement.

When the cell needs energy to perform chemical work, it removes one phosphate group from ATP to yield a less energy rich molecule, *adenosine diphosphate* (ADP), along with usable energy. The energy-releasing phase may be represented by the schematic diagram in Fig. 3–12. Usually, the ADP is reconverted into ATP by energy released from the oxidation of fuels. This is the basic principle of cellular respiration. It can be diagrammed in summary form as in Fig. 3–13.

When the third phosphate group is added onto the ADP using the energy released from the fuel molecule, internal rearrangement of the electron configurations produces the high-energy bond again.

ATP is found in the cells of all living organisms, from bacteria to man. The cell uses ATP as a sort of "money"—a common energy medium which is used to produce almost all the chemical reactions within the organism. Oxidation of fuels builds up a supply of ATP which can be immediately used or temporarily stored, just as having a job builds up a supply of money which can be immediately spent or saved. The money which a man earns is the same in form, no matter what kind of job he has. Similarly, the cell's energy "money" (ATP) is the same, although it may be produced by the oxidation of different fuel molecules. In living organisms, ATP is exactly the same compound, whether found in liver or muscle, fish or fowl.

In summary, then, storing energy in the form of high-energy phosphate bonds has three main advantages. First, it provides a common energy currency for use in any type of biochemical reaction. Second, storing energy in the bonds prevents the release of large amounts of energy which might damage the cell. Third, release of small amounts of energy provides for more efficient use and transformation of energy. In other words, the wastage of energy is greatly reduced.

Thus, the importance of high-energy phosphate bonds to living systems is of the highest order. The universal distribution of such bonds among all living organisms is a further indication of the chemical unity of life.

Chapter 4

The Course and Mechanism of Chemical Reactions

Chemical Reactions and Equations

4–1 INTRODUCTION

This chapter and the next are concerned with two general aspects of chemical reaction: *course* and *mechanism*. The course of chemical reactions refers to overall changes in reactants and products as well as to changes in energy. What atoms or molecules go into and come out of a reaction? Does the reaction release or consume energy as it goes to completion? Study of the course of chemical reactions leads to a consideration of such principles as *reaction rates, free energy changes,* and *chemical equilibria*.

Information gained about the course of a chemical reaction makes possible the drawing of some conclusions about its mechanism. The mechanism of chemical reactions refers to the atomic or molecular interactions by which reactions are thought to occur. To discuss the mechanism of chemical reactions involves such questions as the motion of atoms or molecules, their geometric orientation, and certain environmental effects such as temperature and the influence of catalysts.

4–2 PATTERNS OF CHEMICAL REACTIONS

Chemical equations are like sentences: they contain a complete thought. That thought, of course, does not tell everything that is known about a given reaction. However, it is a useful starting point. Each of the four reaction types discussed below can be recognized by a characteristic pattern. Not all reactions fit these four patterns. However, such a classification is helpful in recognizing the most common forms which chemical reactions may take.

a) Combination reactions. Combination reactions involve the chemical union of two substances to form a third. This may be represented by the generalized equation:

$$A + B \longrightarrow AB \qquad (4\text{–}1)$$

A and B represent two elements or two compounds which react chemically to form the product, AB. As the reaction proceeds, it may either liberate or absorb energy. Two specific examples of combination reactions are shown below. The first, the union of hydrogen and oxygen to form water liberates a great deal of energy:

$$2H_2 + O_2 \longrightarrow 2H_2O \qquad (4\text{–}2)$$

A second reaction, in which oxygen and nitrogen unite to form nitric oxide, requires large amounts of energy to proceed:

$$N_2 + O_2 \longrightarrow 2NO \qquad (4\text{–}3)$$

b) Decomposition reactions. Decomposition reactions involve the breaking of chemical bonds to yield two or more products. They can be written as shown below:

$$AB \longrightarrow A + B \qquad (4\text{–}4)$$

The reactant AB may be a simple two-atom molecule, or a very large one with many atoms. The number of products on the right-hand side of the equation may be more than two. This is often the case with more complex reactants. Decomposition reactions, like combination reactions, may either liberate or absorb energy. For example, the decomposition of water requires energy:

$$2H_2O \xrightarrow[\text{current}]{\text{electric}} 2H_2\uparrow + O_2\uparrow \qquad (4\text{–}5)$$

On the other hand, the breakdown of carbonic acid into carbon dioxide and water, releases energy:

$$H_2CO_3 \rightarrow H_2O + CO_2\uparrow \qquad (4\text{–}6)$$

The upward-pointing arrows indicate products which are given off as gases.

c) Displacement reactions. Displacement reactions occur when one atom or group of atoms replaces another atom or group of atoms in a molecule. Displacement reactions may be written as:

$$A + BC \rightarrow AC + B \qquad (4\text{–}7)$$

Displacement reactions may also involve either the release or the gain of energy. An example of a simple displacement reaction is:

$$Zn + 2HCl \rightarrow ZnCl_2 + H_2\uparrow \qquad (4\text{–}8)$$

Here, the element zinc replaces the hydrogen from the acid molecule, forming zinc chloride. This is an example of an inorganic displacement reaction.

Displacement reactions are important in many reactions within living systems. For example, the iron-containing compound hemoglobin (Hb) combines with carbon dioxide to form a compound known as carbon dioxide hemoglobin ($HbCO_2$). In the lungs, however, oxygen is able to displace CO_2 from its loose combination with hemoglobin:

$$HbCO_2 + O_2 \rightarrow HbO_2 + CO_2 \qquad (4\text{–}9)$$

This reaction aids in the release of carbon dioxide from the blood. At the same time it allows the pickup of oxygen in the lungs.

Displacement reactions among large organic molecules may involve replacing one atom or group of atoms with another. The general reaction plan is similar to the inorganic example given above.

$$\text{Creatine} + \text{ATP} \longrightarrow \text{Creatine-phosphate} + \text{ADP}$$

Here the terminal phosphate of ATP has been transferred to the creatine molecule, giving creatine-phosphate and ADP as products. This particular reaction has an important role in maintaining energy supplies for muscle contraction.

d) Oxidation–reduction reactions. Oxidation is the process by which electrons or hydrogen atoms are removed from a molecule. Reduction is the process of *acquiring* electrons or hydrogen atoms. Oxidation and reduction reactions always occur together. If one molecule is oxidized, another is reduced. Oxidation and reduction reactions are extremely important in biological systems. The general scheme for such reactions is:

$$AH_2 + B \longrightarrow BH_2 + A$$

In this reaction A is oxidized (hydrogen atoms are removed), and B is reduced (hydrogen atoms are added to it).

In biological systems there are specific hydrogen donors and acceptors that are frequently involved in oxidation–reduction reactions. In the reaction below X represents this molecule.

$$\text{Pyruvate} + XH_2 \longrightarrow \text{Lactate} + X$$

In this reaction, pyruvate is reduced to form lactate (hydrogen atoms are added) with the simultaneous oxidation of the molecule XH_2 to form X.

e) Hydrolysis reactions. This is a very important group of reactions all of which involve water as a reactant in breaking a bond (see also Chapter 3, section 3–8). The atoms of the water molecule are then bonded to the products of the reaction.

$$A\text{-}B + H_2O \longrightarrow A\text{-}H + B\text{-}OH$$

Table sugar, sucrose, is composed of two smaller sugars, glucose and fructose. The linkage between these can be broken by hydrolysis.

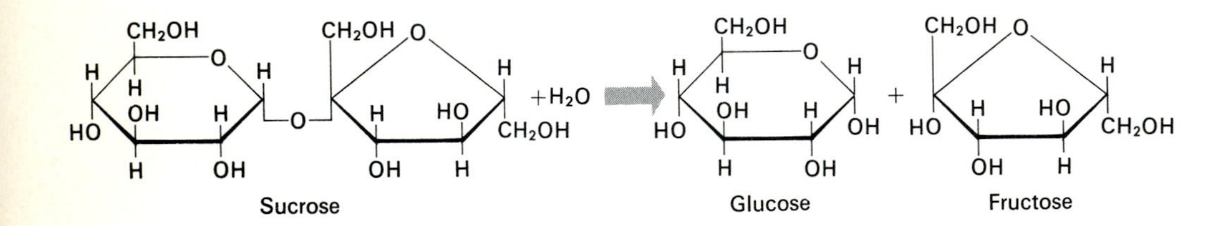

Proteins, nucleic acids, and fats can all be broken down into smaller units by hydrolysis reactions.

All of the reaction types discussed thus far may either release or store energy. This leads to consideration of an important point. Chemical equations *do* inform us of the reactants and products of chemical reactions. They do *not,* however, tell us anything about the energy status of such reactions. This information is obtainable only from experimental data. There is no way to predict, from looking at a chemical equation, whether the reaction releases or stores energy. Since energy is of vital consideration to living systems, it is important to consider some representative reactions from this point of view. This will be the principal point of Chapter 5.

4–3 REVERSIBLE AND IRREVERSIBLE REACTIONS

Chemical reactions are *reversible.* This means that the reaction can go in either direction. In any chemical system, then, two reactions are actually taking place:

$$A + B \rightarrow AB \qquad (4\text{–}12)$$

and the reverse:

$$AB \rightarrow A + B \qquad (4\text{–}13)$$

The forward and reverse equations can be combined into one, with the reversibility indicated by double arrows:

$$A + B \rightleftharpoons AB \qquad (4\text{–}14)$$

Equation (4–14) indicates that at the same time reactants A and B on the left are combining to form product AB on the right, product AB is decomposing to yield the two reactants again. Thus, within one test tube the forward and the reverse reactions are occurring simultaneously.

In Eq. (4–14) the two arrows are of equal length. This shows that the forward reaction occurs just as readily as the reverse. In some reactions, however, this is not the case. Equation 4–15 below shows a longer arrow to the right than to the left.

$$A + B \xrightleftharpoons[\leftarrow]{\quad\quad} AB \qquad (4\text{–}15)$$

This indicates that the reaction occurs more readily in the forward direction than in the reverse direction. Since the forward reaction is favored in this case, we say that the reaction is shifted to the right. The relative size of the two arrows indicates the general direction of reversible reactions.

In principle, all chemical reactions are reversible. There is no reaction known which, under suitable conditions, cannot proceed in the reverse direction. Under ordinary conditions, however, some reactions are far less reversible than others. In these cases, the reaction from right to left occurs so slowly that its rate is barely detectable. For all practical purposes, such reactions are irreversible.

What conditions tend to produce irreversibility? Two factors are important. First, there is the consideration of energy. Some reactions release a great deal of energy going in one direction. Such reactions will tend to go in the reverse direction only if the same amount of energy can be absorbed. An irreversible reaction can be compared

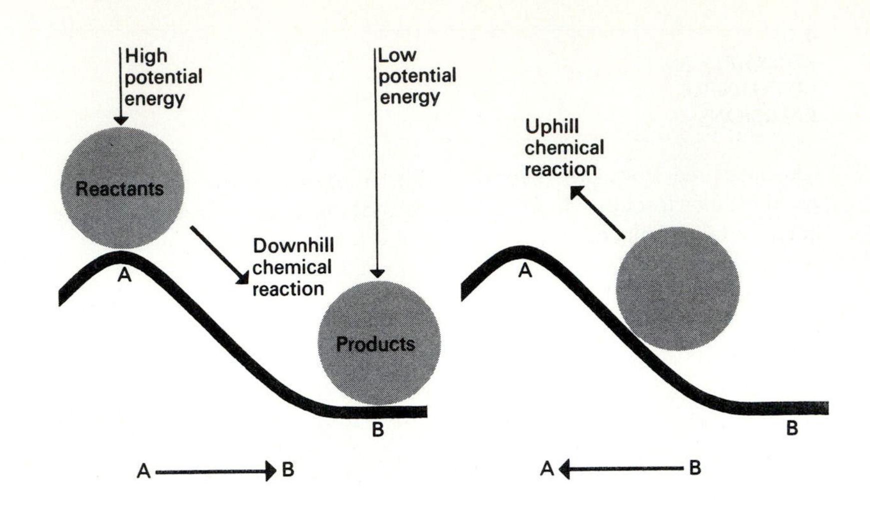

4–1
In this analogy, an energy-releasing chemical reaction is compared to a stone rolling down a hill. When the stone has reached the bottom, it has less potential energy than when at the top. To go in the reverse direction, i.e., back up the hill, the stone will have to absorb the same amount of energy that it released while rolling down. Under natural conditions, absorbing this amount of energy is quite unlikely. Only very rarely, if at all, would a stone ever get back to the top. The same is true of chemical reactions which release large amounts of energy. They are considered to be irreversible. Unless energy is supplied from the outside, the reverse reaction will not occur.

to rolling a stone down a hill. The downward path releases potential energy. To get back up the hill, the stone requires the input of the same amount of energy. More stones roll down a hill than roll back up (Fig. 4–1).

Second, chemical reactions are irreversible if one of the products leaves the site of reaction. This may occur if the product escapes in the form of a gas:

$$A + B \rightarrow C + D\uparrow \qquad (4\text{–}16)$$

or if one product is a *precipitate.* A precipitate is an insoluble substance which settles out of solution:

$$A + B \rightarrow C + D\downarrow \qquad (4\text{–}17)$$

In each case the reverse reaction is inhibited by the removal of one of the product substances (D). Because removal of their products occurs frequently, many biochemical reactions can be considered irreversible.

4–4 BALANCING EQUATIONS AND REACTION PATHWAYS

Not only do chemical equations indicate what reactants and products are involved in a reaction, they also indicate how much of each is involved. Balancing equations shows that chemical reactions conform to the law of conservation of matter. This law holds that in normal chemical and physical reactions, matter is neither created nor destroyed. It is only changed in form. This means that every atom which enters a reaction must be accounted for in the products.

The reaction shown in Eq. (4–18) illustrates the importance of balancing equations. This reaction represents the overall process of *aerobic* (i.e., in the presence of oxygen) *respiration:*

$$C_6H_{12}O_6 + O_2 \longrightarrow CO_2 + H_2O + \text{Energy} \qquad (4\text{—}18)$$

Respiration is a process of energy release which occurs in all living cells.* Equation (4–18) does not indicate the various intermediate reactions, but only the input and output of the total reaction.

However, Eq. (4–18) is incomplete. There are more atoms of carbon, hydrogen and oxygen on the left than on the right. To conform with the law of conservation of matter, every atom must be accounted for. In addition, experimental evidence indicates that the reactants and products of this reaction occur in proportions other than the one-to-one ratios shown above.

By careful measurements, biochemists can determine the amount of glucose, water, and oxygen used and the amount of carbon dioxide, water, and energy released during this reaction. Such measurements indicate that for every molecule of glucose utilized, six molecules of oxygen and six molecules of water are required. This yields six molecules of carbon dioxide and twelve molecules of water among the products. These measurements are always consistent when the reaction is carried out under controlled conditions. Such evidence leads to the conclusion that *chemical reactions occur in definite proportions* and that, under similar conditions, the proportions are always the same.

In order to bring the above equation into agreement with these considerations, it is written in balanced form:

$$C_6H_{12}O_6 + 6O_2 \longrightarrow 6CO_2 + 6H_2O + \text{Energy} \qquad (4\text{—}19)$$

---

* *Respiration* is not to be confused with *breathing*. Breathing is a process whereby organisms get oxygen from the environment into their bodies so that it can be carried to the individual cells. Respiration is the chemical process which occurs within these cells and is responsible for releasing energy to carry out metabolic processes. (See footnote, end Chapter 6.)

Balancing an equation is a means of accounting for all of the atoms of each element during the course of a chemical reaction. It is also a means of indicating the proportions in which reactants or products occur. The numbers used in balancing an equation may represent either large unit quantities, such as a mole, or individual molecules. For example, the 6 in front of CO_2 in Eq. (4–19) can be read as 6 molecules as well as 6 moles of carbon dioxide.

4–5 THE COLLISION THEORY AND ENERGY OF ACTIVATION

No one knows exactly what causes atoms or molecules to interact during a chemical reaction. As a model, the *collision theory* has been of much value in clarifying our ideas about how reactions occur. This theory offers an explanation of how atoms and molecules actually interact. It also helps to explain how factors such as temperature, concentration of reactants, and catalysts affect reaction rates.

The collision theory is derived from the idea that all atoms, molecules, and ions in any system are in constant motion. For any particle to interact chemically with another, both must first come into contact so that electron exchanges or rearrangements are possible. The collision of any two particles is considered to be a completely random event. If two negatively charged particles approach each other, each will mutually repel the other so that direct collision is not likely. The same will be true of two positive particles. If a positive and negative particle approach each other, however, a collision is more likely. Furthermore, this collision may be successful in the sense that it produces an interaction, and hence a chemical change.

Not every collision between oppositely charged particles will produce a chemical interaction. Several other factors are involved. First, the average velocity of the particles determines what percentage of collisions will be successful for any given kind of reactants. The more rapidly the particles travel, the more likely they will yield successful collisions.

Second, particles of each element or compound have their own minimum energy requirements for successful interaction. Assume that we have a system in which molecules of A and B interact to produce C and D. For any collision between A and B to produce a reaction, each molecule must have a certain minimum kinetic energy. We usually speak of this energy in terms of particle velocity. Greater kinetic energy of a particle means greater velocity. Greater velocity means greater probability that a collision will be successful. If the average kinetic energy of a system is increased, the number of successful collisions will, in general, be increased.

Third, molecular geometry plays a role in determining whether a collision is successful. If a molecule collides with an atom or another molecule in such a way that the reactive portion of the molecule is not exposed to the other particle, no reaction will occur. This is true in spite of the fact that the particles may have possessed the proper amount of kinetic energy. For this reason molecular geometry is a factor in any chemical reaction. It is particularly important in reactions between very large molecules. Here, the relative positions of two colliding molecules is crucial to successful interaction. Living systems have developed means of holding large molecules in specific positions which aid in exposing the reactive portion of the molecule. This is one of the main functions of organic catalysts, or *enzymes*.

The minimum kinetic energy required by any system of particles for successful chemical reaction is known as the *activation energy*. Activation energy is a characteristic of any reacting chemical system. If the average energy of the particles is below this minimum, the reaction will proceed slowly, or not at all. If the average is above the minimum, the reaction will proceed more rapidly.

Not all particles are travelling at the same speed within a chemical system. For instance, some of the particles may have no velocity. They may have just collided with a particle of similar charge travelling at the same speed. Others may be moving so rapidly that they pass out of the system completely.

The average velocity of the particles in this system would be expected to lie somewhere between these two extremes. If we plot a graph comparing velocity with number of particles at that velocity, we should get a normal distribution curve. This means that the largest number of particles have velocities somewhere around the average. Such a graph is shown in Fig. 4–2. The dotted line down the center

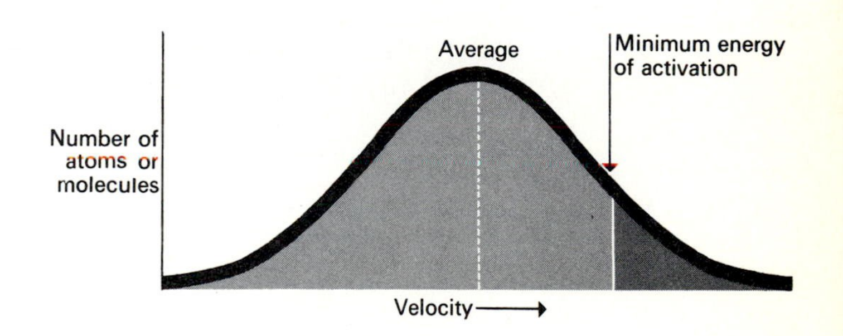

4–2
A bell-shaped curve indicates the distribution of velocities among a large group of atoms or molecules. The largest number of particles show average velocity. Some particles have a low velocity (at extreme left of horizontal axis) while some have a very high velocity (at right). The darker area represents those particles which have enough activation energy to enter into chemical combination.

represents the average velocity. The solid line to the right indicates the minimum energy or velocity requirements for reaction. All particles to the left of this line do not have this energy of activation. Therefore, these particles cannot be expected to react. All particles to the right of this line have velocities greater than the activation energy. They can be expected to produce successful collisions. This graph emphasizes the fact that *chemical interaction between atoms or molecules can be discussed only in terms of probability.*

The rate of a chemical reaction is influenced by factors which increase or decrease the probability that collisions between particles will be successful. This principle is basic to understanding the characteristics of chemical reactions to be discussed in Chapter 5.

Chapter 5

The Course and Mechanism of Chemical Reactions

Energy and Equilibrium

5–1 INTRODUCTION

Of special importance to the biologist is the input or output of energy during chemical reactions. Living things grow and maintain themselves by building molecules. To do this, energy is required. The energy is derived from the chemical breakdown of fuel substances. For this reason, the energy exchanges during chemical reactions must form a major part of the following discussion.

5–2 ENERGY EXCHANGE AND CHEMICAL REACTIONS

All chemical reactions involve an exchange of energy. On the basis of these energy exchanges, chemical reactions can be divided into two classes. Those reactions which absorb more energy than they release are called *endergonic* reactions. Those which release more energy than they absorb are called *exergonic* reactions.

Endergonic and exergonic reactions can be compared in terms of the energy hill analogy discussed in Chapter 4. Endergonic reactions

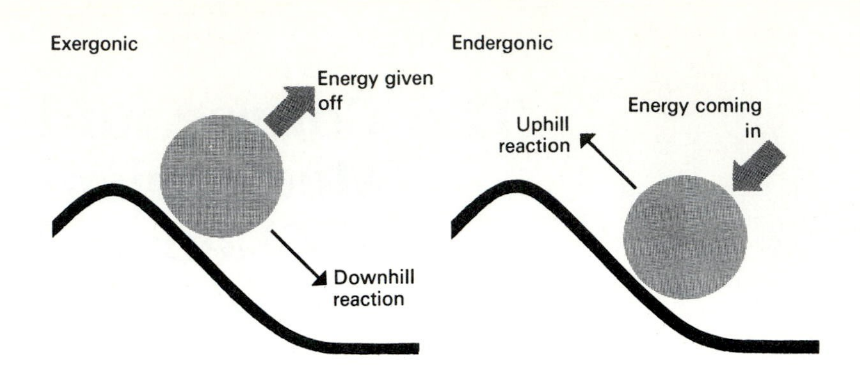

5–1
Exergonic reactions are downhill processes. Hence they release energy. Endergonic reactions are uphill processes and thus require the input of energy. In this analogy, the height of the energy hill represents an amount of energy. The exergonic reaction releases as much energy as the endergonic reaction absorbs.

occur in an uphill direction. Exergonic reactions occur in a downhill direction. This means that, like rolling a stone uphill, endergonic reactions require the input of energy. And, like a stone rolling downhill, exergonic reactions release energy (see Fig. 5–1).

The energy that is available in any particular chemical system for doing useful work is known as *free energy* (symbolized F_0). If a change in free energy occurs during a chemical reaction, the system has either more or less free energy after the reaction than before. The change in free energy of a chemical system is symbolized ΔF_0. An exergonic reaction loses free energy. We can say that such a system shows negative free energy change, $-\Delta F_0$. An endergonic reaction takes in energy. Thus, it shows an increase in free energy, $+\Delta F_0$. It is possible, therefore, to show whether a given reaction involves an overall increase or decrease in free energy simply by putting the symbols $-\Delta F_0$ or $+\Delta F_0$ after the equation.

Like the energy in chemical bonds, energy exchange in reactions is measured in kilocalories per mole of reactant. For example, in the reaction between hydrogen and oxygen to produce water, we find:

$$H_2 + \tfrac{1}{2}O_2 \rightarrow H_2O^* \qquad \Delta F_0 = -56.56 \text{ kcal/mole} \tag{5–1}$$

The reaction has a negative ΔF_0, and hence is one which gives off energy. The more negative the value for ΔF_0, the more energy the reaction releases. Consider the reaction of the sugar glucose and oxygen, which releases, in several steps, the energy for many life processes.

* In equations where energy equivalents are given, the numbers before each molecule refer to number of moles. The one-half O_2 means one-half mole of oxygen.

We can summarize these steps in the following equation:

$$C_6H_{12}O_6 + 6O_2 \rightarrow 6CO_2\uparrow + H_2O \qquad \Delta F_0 = -690 \text{ kcal/mole} \qquad (5\text{–}2)$$

This overall reaction releases a great deal more energy than the reaction shown in Eq. (5–1).

In a similar manner, the numerical value for reactions with a positive ΔF_0 indicates how much energy the reaction requires. In Eq. (5–3), iodine reacts with hydrogen to form the compound hydrogen iodide:

$$\tfrac{1}{2}I_2 + \tfrac{1}{2}H_2 \rightarrow HI \qquad \Delta F_0 = +0.315 \text{ kcal/mole} \qquad (5\text{–}3)$$

This reaction requires a small amount of energy, as shown by the low positive value for the ΔF_0. On the other hand, the process of photosynthesis, in which green plants produce carbohydrates from carbon dioxide and water, requires a large intake of energy. This energy is supplied by light. The overall process can be written as:

$$6CO_2 + 12H_2O \rightarrow C_6H_{12}O_6 + 6H_2O + 6CO_2\uparrow \qquad \Delta F_0 = +690 \text{ kcal/mole} \qquad (5\text{–}4)$$

Knowing the ΔF_0 makes it possible to compare the amounts of energy which various reactions absorb or release.

All exergonic reactions show an overall loss of free energy. Many of these same reactions, however, require an energy input to get them started. If left to themselves, many reactants will never show any chemical activity. However, if the right amount of energy is supplied, the reaction begins. It then goes to completion without the addition of more energy from the outside.

How can this be explained? In such chemical systems, the reactants have relatively high activation energies. The addition of energy gets a larger percentage of particles in the system up to the required kinetic energy. In absorbing this energy, the particle becomes activated. When particles are in an activated state, a successful reaction is much more probable.

A specific example will clarify this point. Formic acid, HCOOH, is the pain-causing substance in wasp and bee stings. Under certain conditions, formic acid decomposes into carbon monoxide (CO) and water, with a slightly positive ΔF_0. However, for this reaction to occur, a formic acid molecule must first absorb enough energy to become activated. In being activated, the molecule undergoes a rearrangement of one hydrogen atom. The molecular structure is changed, and along with it the stability of the whole molecule. It splits into two parts, carbon monoxide and water:

$$HCOOH \rightarrow CO + H_2O \qquad (5\text{–}5)$$

Energy exchanges in chemical systems are often given on a graph, which shows the changes in potential energy during the course of reaction. These changes are then compared with the time it takes the reaction to go to completion. Such a graph for the formic acid reaction is shown in Fig. 5–2.

Analysis of this graph shows some important things about this chemical reaction. The graph describes the changes in energy for one molecule as that molecule undergoes the decomposition reaction shown in Eq. (5–5). Before reacting, an individual molecule is in a relatively low energy state. By absorbing energy, this molecule passes to a higher potential energy level. It is now in an activated state. The appropriate rearrangement occurs, and the product molecules are formed.

Note that the product molecules are at a slightly higher energy state than the original molecule of formic acid. This indicates that the overall reaction absorbed a small amount of energy. The reaction is endergonic.

The distance *h* on the graph indicates the energy of activation for this chemical system. The height of the graph line can thus be considered an *energy barrier:* a "hill" over which the molecule has to climb before it can roll down the other side to completion. After absorbing the required activation energy, the reaction proceeds spontaneously, just as a stone rolls down a hill once it is pushed over a rise at the top.

Figure 5–3 shows an energy diagram for an exergonic reaction. The products of exergonic reactions are always at lower potential energy states than the reactants.

Under standard conditions* hydrogen and oxygen exist together in a single container without the least indication of reacting to produce water. The molecules simply do not have the necessary energy of activation. However, if a small electric spark is introduced into the chamber, an explosive reaction takes place, releasing a great deal of energy. Thus, hydrogen and oxygen react quickly to form water if given the necessary push to get them started. The fact that a spark is all that is needed to provide this push shows that the energy barrier for this reaction is not very high. Passing the spark through hydrogen and oxygen provides enough energy to put some molecules of each in the activated state. The activated molecules spontaneously react to form molecules of water. From Eq. (5–1) we see that this reaction releases energy. The energy released from one reaction is enough to get several other molecules of each element over the energy barrier. The spark provides an initial push. The rest of the reaction occurs by a chain-reaction effect, like a house of cards which collapses when one card is disturbed.

---

* *Standard conditions* means standard temperature (0° on the celsius or centigrade scale) and one atmosphere of pressure (760 millimeters of mercury).

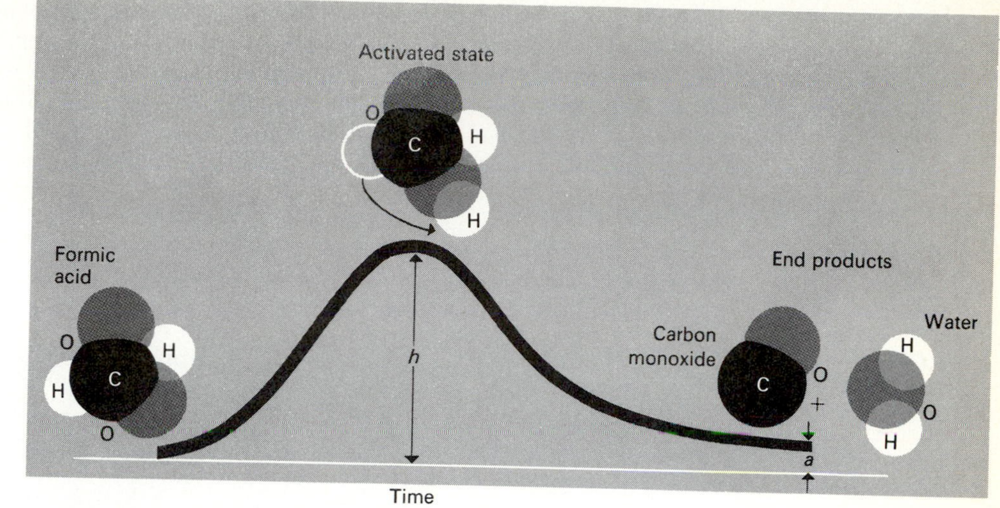

5–2
Changes in potential energy during the decomposition of formic acid. The original molecule, to the left, is in a relatively low energy state. By collision with another molecule of high kinetic energy, this molecule becomes activated. The potential energy of such an activated molecule is greater. During activation, a molecular rearrangement occurs and the molecule splits. The end products, carbon monoxide and water, are at a higher energy state, *a* on the graph, than the original molecule. Distance *h* represents the height of the energy barrier.

H
O
O
H
Potential energy
h
H
H
O
+
O
Time

5–3
A spontaneous exergonic reaction, the formation of water from H_2 and O_2. The end products are at a lower energy state than the two starting reactants, indicating that the reaction releases energy. This reaction has a small energy barrier, as indicated by the size of *h*.

From the point of view of energy, the two most important chemical processes in biology are those of photosynthesis and respiration. The photosynthetic reaction is an endergonic one. By using solar energy, green plant cells run an uphill reaction. In the course of this reaction, many chemical bonds are built and broken. However, the energy states of the end products are higher than those of the reactants.

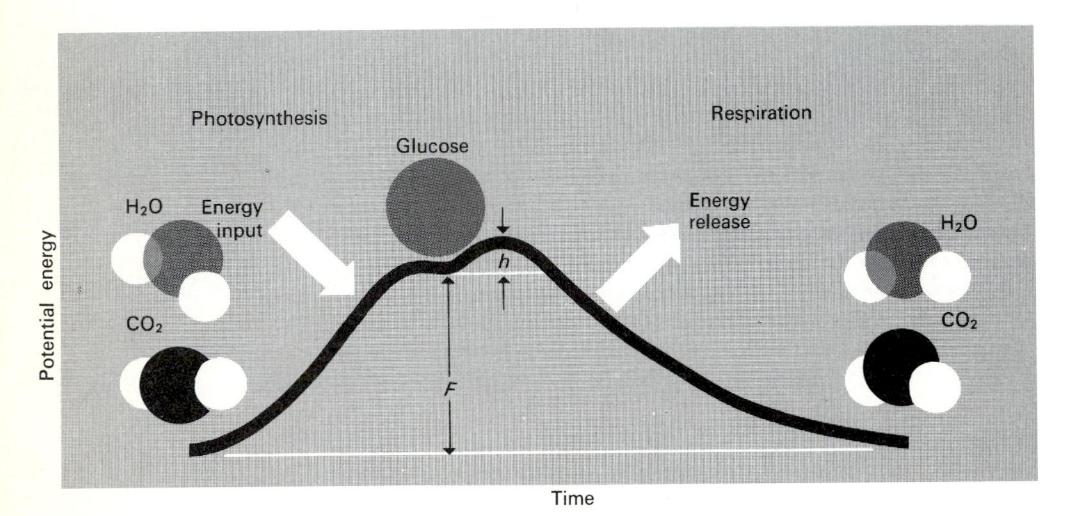

5–4
The beginning reactants for photosynthesis are carbon dioxide and water. Energy from sunlight drives this uphill reaction to produce a food molecule, glucose. Respiration involves the breakdown of glucose and other fuel molecules to produce energy. The end products of this downhill reaction are carbon dioxide and water. The energy of activation necessary to begin the respiratory process is represented by h. The total difference in free energy between glucose and the end products is represented by F.

Respiration is an exergonic reaction. It results in the release of the potential energy stored in a food molecule. Whereas the photosynthetic reaction goes uphill, the respiratory reaction goes downhill. The relationship between these two complementary reactions is shown in Fig. 5–4. Molecules of a food rest at the top of the "energy hill." In order to go down the hill, they must pass over the small energy barrier, h. The distance which the molecule moves down the right side of the graph represents the loss of free energy.

The following statements summarize this discussion of energy and chemical reactions.

1. Endergonic reactions are those which absorb more energy than they release. Exergonic reactions release more energy than they absorb.
2. Endergonic reactions show an increase in free energy, $+\Delta F$. Exergonic reactions show an overall decrease in free energy, $-\Delta F$.
3. The end products of endergonic reactions have a higher potential energy state (possess greater free energy) than the end products of exergonic reactions.
4. Energy of activation is the amount of energy required to put an atom or molecule into an activated state. This means that the atom or molecule possesses a higher potential energy.
5. The amount of energy of activation for any given chemical system is a characteristic of that system. The amount of activation energy does not necessarily determine whether the overall reaction is endergonic or exergonic.
6. Combination and decomposition reactions can be either endergonic or exergonic.

5–3 RATES OF REACTION

The *rate* of any chemical reaction is defined as *the amount of reaction in a given period of time.* The amount of reaction is generally measured in terms of the change in concentration of reactants or products. The basic relationship between amount of reaction and time can be expressed as a word equation:

$$\text{Rate of reaction} = \frac{\text{Change in concentration}}{\text{Change in time}} \qquad (5\text{–}6)$$

The concept of rate in chemical reactions is vital to an understanding of chemical equilibrium. In addition, knowledge about reaction rates allows a clearer understanding of the mechanisms by which a particular chemical process occurs. Under given conditions, the rate of a reaction is a predictable characteristic of chemical systems. Such systems can thus be described in terms of reaction rates, as well as direction or energy exchange.

A specific example will help in understanding the concept of rate as it applies to chemical reactions. Consider the reaction below, where molecules of A and B combine to yield the products C + D.

$$A + B \rightharpoonup C + D \qquad (5\text{–}7)$$

If we begin with molecules of A and B only, the rate of reaction at the outset is very high. Rate in this case can be measured in terms of how rapidly the reactant molecules A and B disappear in specific units of time. If we were to stop the reaction every thirty seconds and determine the amount of A and B present, these data would, plotted on a graph, give a line like that shown in Fig 5–5(a).

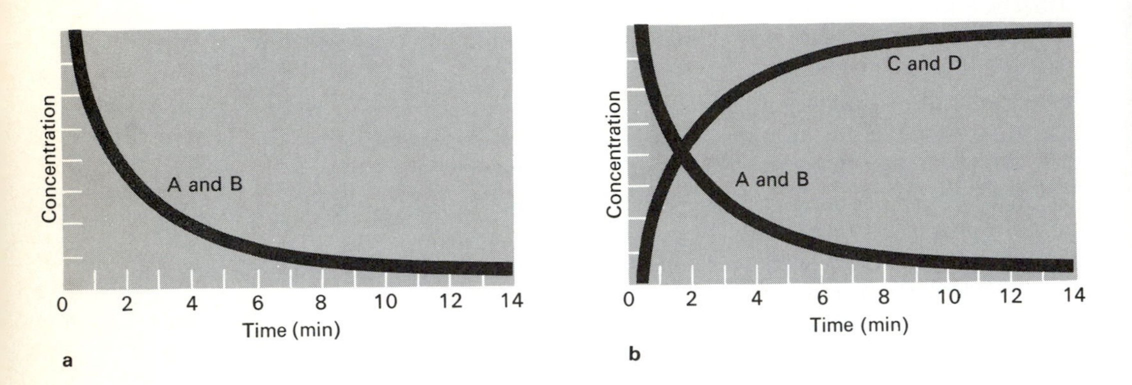

5–5
Changes in concentration of reactants A and B and products C and D during the course of a chemical reaction. This relationship between change of concentration and time indicates the rate of reaction. Graph (a) shows only change in concentration of reactants. Graph (b) also shows the change in concentration of products. Change in concentration of either reactants or products is most rapid during the first few minutes of reaction.

We can see that the rate at which A and B disappear from solution changes with time. For instance, the concentration of A and B decreases most rapidly in the first two minutes and begins to level off about the sixth minute. By the ninth minute nearly all of A and B have been used up in the reaction. At this point, the rate at which A and B combine to yield C and D is almost zero.

If we also plot the rate of appearance of products C and D in this reaction, a curve is obtained which is just the opposite of that for the disappearance of A and B, Fig. 5–5(b). This is not surprising, since the rate at which C and D appear depends directly upon the rate at which A and B interact.

The molecular explanation for the change in rate of chemical reactions goes back to the collision theory. The rate at which a chemical reaction progresses toward completion depends upon the number of effective collisions between reacting molecules or atoms. This number is determined for any given reaction by the concentration of reactants. The more molecules or atoms of reactants, the greater the number of effective collisions.

During the course of any chemical reaction the concentration of reactants (in a closed system) continually decreases. This reduces the chances of a collision between a molecule of A and a molecule of B. At the same time, the concentration of product molecules C and D is increasing. This means that collisions between C and D molecules will become more frequent. If such collisions produce a reaction to yield A and B again, then the reaction is a reversible one. If C and D do not interact in this way, then the reaction is irreversible. The reaction graphed in Fig. 5–5 can be seen to be irreversible, since the concentration of reactants steadily decreases and that of products steadily increases.

An analogy may clarify this situation. Many fairs have enclosed arenas where small rubber-bumpered automobiles can be driven about. The cars are generally powered by an electric motor. Each car holds one rider and is free to move anywhere within the boundary of the arena. Let us suppose that there are equal numbers of red and blue cars in the arena at a given time. Let us further suppose that every time a red car collides with a blue car, either from the back or from the front, their bumpers lock. At the outset, the cars move about the arena in a random fashion. The chances that a blue car will collide with a red car are 50 : 50. That is, it can be assumed that 50 percent of the collisions will occur between red and blue cars. Some red-blue collisions will produce effective bumper locks. These cars, therefore, are no longer single red or blue vehicles. They are now locked together as two-car units.

It is apparent that as time goes on, the number of free red and blue cars will become fewer and fewer. As a result, the interaction between red and blue will be less frequent. The same principle applies to the chemical reaction discussed in Eq. (5–7).

5–4 FACTORS INFLUENCING RATES OF CHEMICAL REACTIONS

There are many factors which influence the rate of chemical reactions. Each of these factors influences the rate at which collisions occur between molecules or atoms. A change of conditions alters the rate of chemical reactions by increasing or decreasing the number of effective collisions. The mechanism by which each factor acts on collision rate will be discussed below.

(a) The effect of temperature. Chemical reactions are quite sensitive to changes in temperature. An increase in temperature increases the rate of reaction. A decrease in temperature decreases the rate of the reaction.

Temperature affects reaction rates by changing the average velocity of particles in a chemical system. Increasing the temperature increases particle velocity. Decreasing the temperature produces the opposite effect. Change in velocity means two things. First, if the speed of a particle is increased, the number of collisions between atoms or molecules is increased. This, in turn, increases the number of effective collisions in a given unit of time. Second, increasing the velocity of a particle also increases its kinetic energy. This means that each collision between reactant particles has a better chance of being successful. An increase in temperature thus pushes all the particles toward an activated state. A decrease in temperature works in the opposite manner.

Pressure has the same effects as temperature on reaction rates. An increase in pressure within any chemical system increases the density. Since the particles are pushed closer together, the number of collisions is increased.

(b) Concentration of reactants. In Section 5–3 we saw that the rate of reaction $A + B \rightarrow C + D$ decreased with time because molecules of A and B were gradually used up. This means that the concentration of A and B decreased. This indicates the dependency of the rate of reaction on the concentration of reactants. Doubling the concentration of A and B would increase reaction rate by a factor of four. Doubling only the concentration of one reactant (let us say, A) would increase the rate of reaction by a factor of two. This can also be understood in light of the collision theory.

By doubling the initial concentration of only substance A, we have, nevertheless, doubled the chances of a collision between A and B. Each molecule of B now has twice as great a chance of colliding with a molecule of A. This doubles the rate at which effective collisions occur. Note that doubling the concentration of A alone will not increase the total amount of products. The number of molecules of C and D at the end of the reaction is determined by the available amount of the least-concentrated reactant, in this case substance B.

Doubling the concentration of both reactants increases both the rate of reaction as well as the total amount of end product. Every molecule of A now has twice as many chances of colliding with a molecule of B. In turn, every molecule of B has twice as many chances of colliding with a molecule of A. As a result, molecules of A and B will collide four times more frequently. The number of effective collisions will therefore be four times as great. At the same time, the total amount of product at completion is doubled.

This situation is analogous to building a brick house. The two necessary components for such a structure are bricks and mortar. The total size of the house at the end of construction is determined by the availability of both components. Bricks are of no value without mortar, nor mortar without bricks. Whichever is used up first, the result is the same: construction comes to a halt. If the supply of bricks is plentiful, the rate of construction will be increased by increasing the availability of mortar. Likewise, if mortar is plentiful, increase in availability of bricks will increase rate. In no case, however, can the total construction exceed the available number of bricks or the amount of mortar.

Changes in the concentration of products in irreversible reactions generally have no measurable effect on reaction rate. However, a change in the concentration of products in a reversible reaction has a very noticeable and important effect. This effect is more closely related to the topic of chemical equilibrium. It will be discussed in some detail in Section 5–6.

(c) Degree of subdivision of reactant particles. In order for chemical reactions to occur, the atoms or molecules of the reactants must collide in such a way that electron interactions are possible. If matter is divided into fine enough particles, individual atoms or molecules can interact. For this reason, many reactions take place in a solvent. Molecules of the solvent break up the solute into parts which are small enough to interact chemically with particles of other substances. Degree of subdivision therefore indicates the number of particles of reactant exposed.

An example of the chemical effect of subdivision is found in the digestion of fats. Fats often occur in large globules which expose relatively little surface area. Bile, from the gall bladder, breaks the fat globules into very small droplets. This process is called *emulsification.* The droplets of emulsified fats display a far greater surface area than the original globules. Hence, a greater number of fat molecules are exposed to the action of fat-digesting enzymes. The rate at which the fat molecules are broken down is greatly increased.

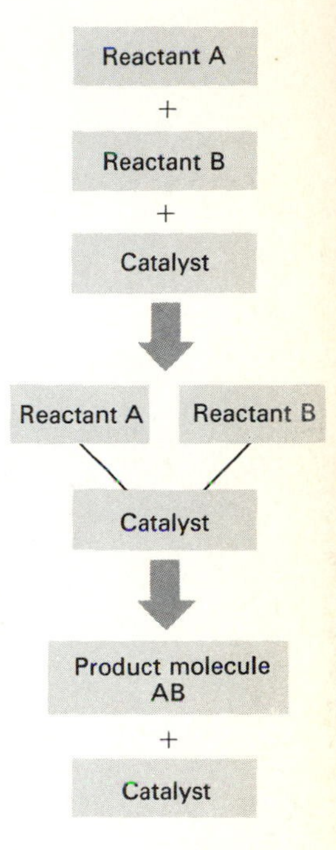

5–6
A catalyst hastens the rate of chemical reactions without being used up in the reaction. The reactants combine with the catalyst to form a temporary complex (middle portion of figure). This breaks down to release the product molecule AB and the catalyst, which is then ready to take part in another such reaction.

(d) The effect of catalysts. Catalysts affect the rate at which chemical reactions occur. Figure 5–6 illustrates the pattern in which catalysts operate. Molecules A and B interact with the catalyst to form a complex. This complex breaks down to yield the product molecule AB and the catalyst. Thus, *the catalyst can be used again in another reaction.* The chemical structure of the catalyst molecule is not permanently changed during such reactions. A catalyst cannot make an interaction occur which would not occur of its own accord without the catalyst. Its function is simply to increase the rate at which such reactions take place. Those catalysts of greatest importance to living organisms, enzymes, will be discussed in detail in Chapter 10.

5–5 CHEMICAL EQUILIBRIUM

Within a certain period of time, reversible reactions reach a state of equilibrium. When this condition is reached, the *proportion* of reactants in relation to products remains the same. Note this does *not* mean the *amounts* of reactant and of product are necessarily equal.

An example will illustrate this point. Consider the reaction in which molecules A and B yield products C and D:

$$A + B \rightleftharpoons C + D \qquad (5\text{–}8)$$

The different lengths of the arrows indicate that the conversion of A and B to C and D occurs more readily than the conversion of C and D to A and B. In other words, the energy barrier is lower for a successful interaction of A and B than for C and D. This means that, in a given period of time and at the same concentrations, more A and B will react with each other than will C and D. Under the same conditions, the initial rates of the two reactions are different.

If the reaction begins with only molecules of A and B present, it will occur at first only to the right. The graph in Fig. 5–7 plots change in concentration of reactants and products. It is apparent that the change in concentration ceases after about the fifth minute. Beyond this point, there is no further change. Further, when these curves level off, the concentration of C and D is greater than the concentration of A and B. The reaction is thus directed to the right, as the relative lengths of the arrows indicate. The energy of activation for the forward reaction is less than that for the reverse reaction.

As long as molecules of reactant and product are still present in a system, chemical reactions never cease. In the above reaction, C and D accumulate because of a relatively high activation energy. This means that eventually the number of collisions between molecules of C and D will be higher than the number of collisions between molecules of A and B. As a result, the rate of the reverse reaction will increase, despite a higher energy of activation which tends to oppose it. At the same time, the rate of the forward reaction will decrease, due to the decreasing concentrations of A and B.

Eventually, a point will be reached at which the forward rate equals the reverse rate. When this condition is reached, we say that a state of *chemical equilibrium* exists. Note that the concentration of reactants does *not* have to equal the concentration of products in order for equilibrium to be established. In the above case, the concentrations of C and D are much greater than the concentration of either A or B at the equilibrium point. The important feature of equilibrium is that *the rates of forward and reverse reactions are the same.* This means that as many molecules of A and B are being converted into C and D as molecules of C and D are being converted to A and B.

An analogy may help to understand how chemical equilibrium is established. In Fig. 5–8 two boxes are shown, side by side. Between the boxes is a partition, somewhat lower than the other walls. Into box 1 are placed about 20 active frogs. For purposes of this analogy, we shall assume that these frogs jump in a completely random fashion. From time to time, a frog will jump just high enough so that it

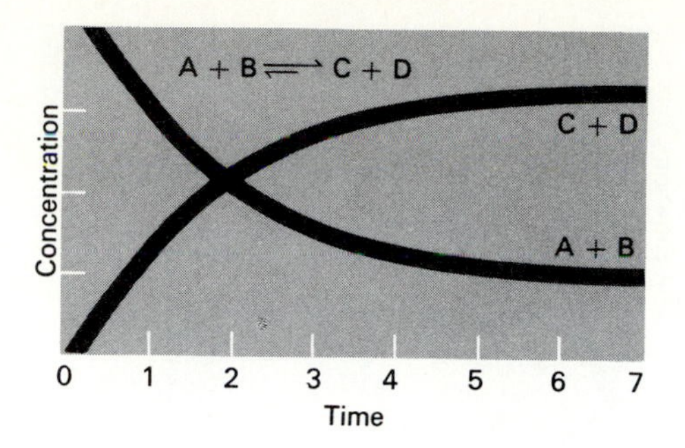

5–7
Graph showing the change in concentration of reactants and products in a reversible reaction. The point of chemical equilibrium is reached where the two curves level off (at about the fifth minute). Here, the rate of the forward reaction is equal to the rate of the reverse reaction.

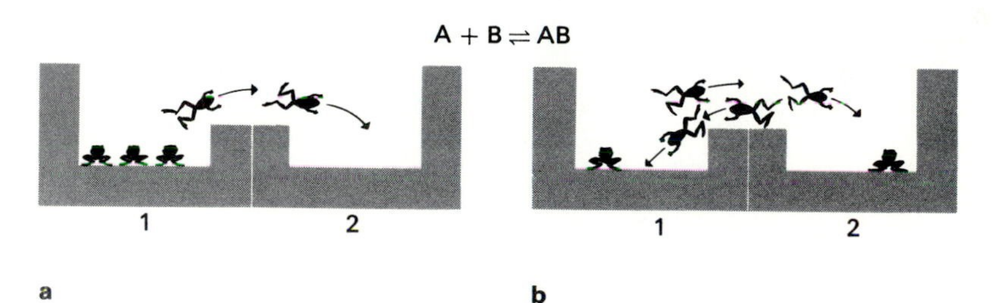

5–8
Illustration of the establishment of chemical equilibrium with jumping frogs. Drawing (a) shows the start of the "reaction" with all the frogs in one compartment. Occasionally one clears the barrier and passes into the next chamber. Eventually an equilibrium is established when the number of frogs jumping in one direction equals the number jumping in the other direction. The frog system is said to be in a state of dynamic equilibrium.

can pass over the partition separating box 1 from box 2. It will therefore land in the second box.

Since all the frogs start out in box 1, any frog which hops in the right direction with enough energy to clear the partition will pass from box 1 into box 2. We will call this "the forward reaction." As time goes on, however, the number of frogs in box 1 decreases, while that in box 2 increases. Those frogs now in box 2 are also hopping about, and from time to time one of them will jump over the partition back into box 1. It follows that the rate (i.e., the number of frogs passing over the partition in a given unit of time) at which frogs jump back into box 1 will depend partly upon how many frogs there are in box 2. The rate at which frogs jump from 1 into 2 will be very great

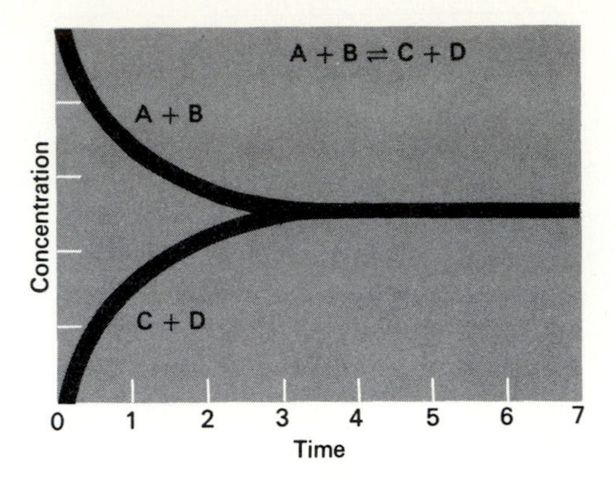

5–9
A graph showing changes in concentration over a period of time for a completely reversible reaction. Note that an equilibrium is established about 3½ minutes after the reaction has begun. At this point, concentration of reactants equals concentration of products.

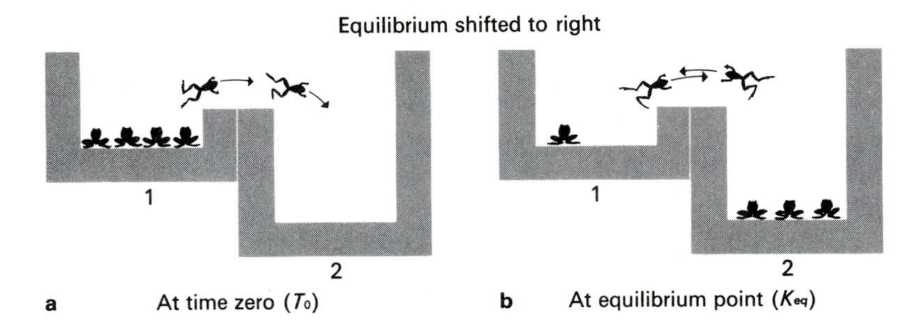

5-10
A jumping-frog model with one box lower than the other represents (a) a system analogous to a chemical reaction with equilibrium directed to the right. More frogs accumulate in chamber 2 because the barrier back into chamber is now higher than from 1 into 2. (b) Equilibrium is again established when the rate at which frogs jump in one direction is the same as the rate at which they jump in the other.

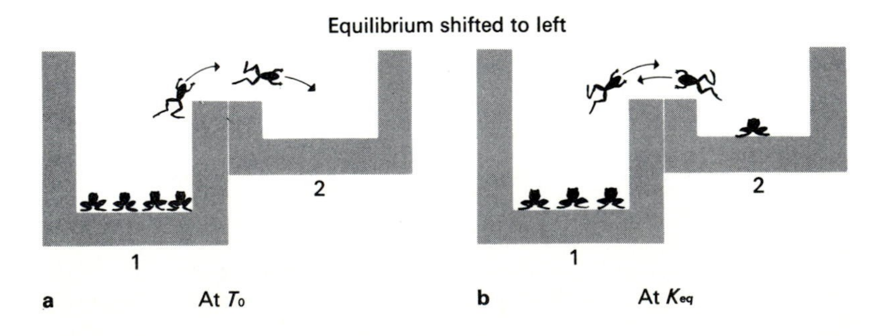

5–11
Jumping frog model for a reaction in which equilibrium is directed to the left. In this reaction, more product accumulates on the left than on the right. At equilibrium, however, the rate of reaction to the right equals the rate of reaction to the left.

at first, since all the frogs are in box 1. As frogs accumulate in box 2, however, the rate of reverse jumps will steadily increase. Eventually, a point will be reached at which the forward rate will equal the reverse rate. This represents the equilibrium point for the frog system, Fig. 5–8(b).

In Fig. 5–8, equilibrium is established at a point where the number of frogs in each compartment is the same. This results from the fact that an individual frog may jump over the partition as easily from box 1 as from box 2. In either case, the height of the jump is the same. Making the jump is analogous to atoms or molecules becoming "activated." The higher the partition, the fewer frogs that jump over in a given period of time. In chemical terms, the higher the energy of activation, the fewer molecules that are activated in any period of time. But since the height of the barrier is the same in either direction, the rate of forward or reverse jumps will be determined only by the number of frogs in each compartment.

As we saw in Chapter 4, p. 59, some reactions are completely reversible. Such reactions are symbolized by putting arrows of equal lengths pointing in both directions:

$$A + B \rightleftharpoons C + D \tag{5–9}$$

A graph showing rates of reaction for completely reversible systems can be seen in Fig. 5–9. The converging of the two lines indicates that the concentration of reactants and products is equal after the reaction reaches equilibrium.

The shifting of equilibrium to either the right or left can be illustrated by the jumping frog systems shown in Figs. 5–10 and 5–11. Figure 5–10 shows an equilibrium directed to the right. Since the barrier between compartments 1 and 2 is much lower in moving to the right, more frogs will be able to jump in that direction. Jumping back from compartment 2 to 1 is much more difficult. Hence at equilibrium more frogs will be found in compartment 2 than 1. In chemical terms, the energy barrier is much greater going to the left than to the right. Thus the equilibrium point favors the forward reaction. The reverse is true of the system in Fig. 5–11.

When a point of equilibrium is reached in the frog system, the passage of individual frogs across the partition *has not ceased.* The passage continues as long as the frogs continue jumping. The important feature is that *the rate of jumping in one direction is equal to the rate of jumping in the other.* The same is true of chemical reactions at equilibrium point. The reaction is still occurring both to the right and to the left. Since the rates of these reactions are equal, however, the concentration of reactants and products remains the same.

The condition which exists at chemical equilibrium is referred to as a *dynamic equilibrium.* While the overall characteristics of a given system at dynamic equilibrium remain relatively constant, its individual parts are in a state of continual change.

Left to itself, the direction which any reaction takes is toward a condition of equilibrium. The point at which chemical equilibrium lies is characteristic for any given chemical reaction.

What happens when a chemical reaction at equilibrium is disturbed by the removal or addition of substances on either side? Suppose that the equation

$$A + B \rightleftharpoons AB \tag{5–10}$$

exists in perfect equilibrium. If we add either substance A or B, or both simultaneously, we will push the reaction to the right so that more molecules of AB will be formed. We can accomplish the same effect by another method. Without adding more of either A or B, the reaction may be shifted to the right by removal of some AB. If we add molecules of AB to the system in equilibrium, the reaction is shifted to the left. The equilibrium point in such reactions has not been changed by adding or removing substances. Only the direction or rate of the reaction has been changed momentarily. Reversible reactions eventually return to dynamic equilibrium.

How does all this apply to living organisms? Nearly all known biochemical reactions are reversible. Hence, they exist in a state of dynamic equilibrium. But a living system has need for the continual, one-way production and breakdown of molecules in order to carry out its activities. Were all metabolic reactions to attain equilibrium, death would result. Living organisms manage to decompose and synthesize organic molecules at extremely rapid rates, despite the fact that all the reactions are reversible. Their task is further complicated by the fact that biochemical reactions generally take place in a series of steps. This means, from original reactant to final product, a number of intermediate reactions and products are involved. The process by which sugars are broken down into carbon dioxide and water, for example, involves over thirty individual chemical reactions. Nearly every one of these reactions is reversible.

The question now becomes: How is a large series of reversible reactions kept going in one direction within a living organism? The overall process goes in one direction because *the products from each individual reaction are used as reactants for the next reaction in the series*. This can be represented in the generalized scheme:

$$\overrightarrow{A \rightleftharpoons B \rightleftharpoons C \rightleftharpoons D \rightleftharpoons E} \quad \text{(End product)} \tag{5–11}$$

Substance A may be considered the raw material which the cell takes in from the outside, and substance E the end product. The individual reactions are completely reversible. Yet, the arrow indicates that there is a one-way direction for the series.

Assume that the end product E is utilized in some cell activity. As soon as it is produced, it becomes involved in another series of reactions. This means E cannot go back in the opposite direction to produce D. Thus, the continual removal of E from the reaction site keeps the reaction series shifted to the right. In addition, the concentration of substance A within the cell may be increased by continual addition from the outside. This adds a push to the right. The result is a precisely controlled set of interactions producing the overall effect of a one-way reaction series.

Such a situation exists in respiration. The overall process of respiration involves the breakdown of glucose to yield carbon dioxide, water, and usable metabolic energy. This is actually a many-stepped reaction which may be diagrammed as:

$$C_6H_{12}O_6 + 6O_2 \rightleftharpoons A \rightleftharpoons B \rightleftharpoons C \rightleftharpoons D \cdots \rightleftharpoons 6CO_2 + 6H_2O \qquad (5\text{—}12)$$

All of the many intermediate reactions in this process are in themselves reversible. In respiring cells, however, the carbon dioxide diffuses through the cell membrane and is removed. It thus leaves the reaction site. As a result, the reaction series shows an overall movement to the right, despite the reversibility of the individual reactions in the series.

5–6 ENTROPY AND FREE ENERGY

We have seen that energy changes are of great importance as a driving force in many chemical reactions. We have also seen that, in general, chemical reactions tend to proceed from higher to lower energy states. In other words, a chemical reaction can be visualized as proceeding down the energy hill, eventually reaching a point where the slope of the hill levels off.

However, some chemical reactions seem to roll *up* the energy hill in going to completion. It is apparent that such reactions are driven by something other than the tendency to proceed from higher to lower energy states. These and other considerations have led to the development of the *Second Law of Thermodynamics.* It is useful to consider biological processes in light of this generalization.

The Second Law of Thermodynamics relates changes in the free energy of a chemical system with changes in *organization* of the parts of that system. The concept of *entropy* is used in physics and chemistry to represent the amount of *disorganization* which any system shows.

Consider a room filled with molecules of one kind of gas. We can view the distribution of the gas molecules from a standpoint of probability. It is highly improbable that all of the molecules would be located in one corner of the room. Instead, it is much more probable that these molecules will be equally distributed. In other words, there are no more molecules in one area of the room than in another.

If all the molecules of gas are located in one corner of the room, the degree of organization in this system is rather high. Recall that the molecules of any substance are constantly in motion. Left to themselves, they tend to spread out, distributing themselves uniformly over the entire space. Thus, it is highly improbable that gas molecules will accumulate in the corner of a container by their own random motions. It would be necessary to make an organized effort, that is, expend free energy to force all the molecules into one area. If the attempt were made, it would bring about an increase in the organization of the system.

If the system is so organized and then left to itself, the molecules begin scattering about the rest of the container. The system begins to become more disorganized almost immediately. It is characteristic of both chemical and physical systems that, left to themselves, they tend to become more and more disorganized.

Everyone is familiar with this principle from day-to-day experience. It is well known that if a house is not cleaned and kept in repair, it tends to become more messy. We might say that it becomes more disorganized. A constant expenditure of energy is necessary to keep things from becoming disorganized. A human being is a very complex organization of specific parts. Yet, that organization can be maintained only as long as a certain amount of energy is expended to overcome the tendency to assume a more random state. The energy for maintaining this order comes from the food we eat.

We are now in a position to understand what entropy means. Since it is a numerical measure of disorder, entropy can be defined in mathematical terms and given the symbol S. This definition states that the less probable a given distribution of molecules or atoms in a system, the less entropy which that system contains. The greater the disorganization, the greater the value for S; the more organization (i.e., the more order), the less the value for S. In other words, the more probable a given distribution, the greater the entropy in that system.

It follows, therefore, that physical or chemical systems can do work as they proceed from a state of low entropy to a state of high entropy. For example, as the molecules of gas in a room move from one corner into the rest of the room, the pressure which their diffusion exerts can be used to move matter. Consider the box shown in Fig. 5–12. The only way that a molecule of gas can get from one compartment to another is to go through the opening shown in the center partition. This opening is blocked by a paddle-wheel device. As molecules hit the blades of the paddle-wheel, they exert enough pressure to turn the wheel.

In diagram (a), compartment 1 contains more molecules than compartment 2. Since they are both parts of the same system, compartment 1 has a greater degree of organization than compartment 2. Therefore, the entropy of compartment 1 is less than that of compartment 2. Another way of saying this is that one portion of the

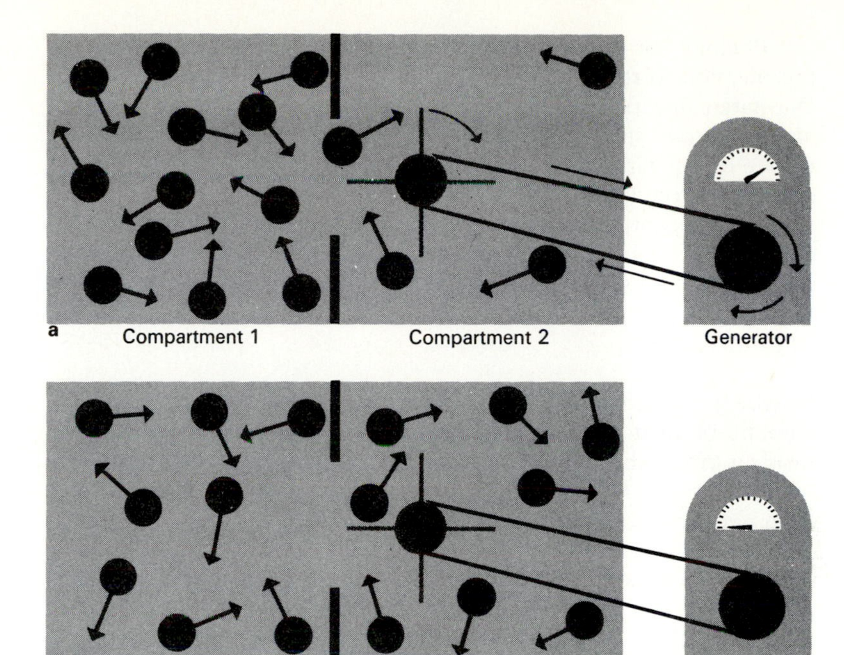

5–12
The closed box (a) has two compartments containing a gas, and represents one complete physical system. There are more molecules in compartment 1 than in compartment 2. This represents an organized state, since gases tend to diffuse evenly throughout a container until a completely random distribution is achieved. Thus there is an *increase* in entropy. While this happens, the system performs work by turning the wheel connected to an electric generator. (b) When the difference in energy levels between two parts of the system disappears, the system no longer performs work.

system (compartment 1) has less entropy than another portion of the system (compartment 2). As a result, there will be greater movement of molecules from compartment 1 into compartment 2. This will turn the paddle-wheel to the right. This, in turn, will run the generator, doing work.

In time, however, the molecules will become evenly distributed on each side of the partition, as shown in diagram (b). The level of organization in each compartment is therefore the same. Under these conditions, there is no difference in the entropy of the two parts of this system. Hence, there is no greater movement in one direction than in another. The system is at equilibrium and cannot yield any more work.

The driving force behind this reaction is the tendency of the system to proceed from a more organized to a less organized state. In other words, *the driving force is toward an increase in entropy*.

Chemical or physical reactions proceed only from states of high organization to states of low organization, unless energy is supplied. Or, stated in more precise terms, *reactions proceed from lower to higher entropy states*. The relationship between entropy values and degree of organization is thus an *inverse* one. The higher the degree of organization, the lower the entropy, and vice versa.

The concept of entropy is also related to that of free energy, discussed in Section 5–2. In terms of chemical reactions, this relationship can best be shown by a simple equation:

$$\Delta F_0 = \Delta H - T\,\Delta S \qquad (5\text{–}13)$$

In words, the change in free energy of a system is equal to the amount of heat given off or absorbed by the system minus the product of the absolute temperature and the change in entropy. This relationship may look complex. In reality, however, it follows directly from our earlier considerations of free energy.

Why do heat and temperature (the measure of heat) appear in this equation? Recall that heat involves the random motion of molecules. In a very real sense, heat is chaotic energy. It is obvious that heat will have a very great effect upon the order or disorder of a system, i.e., upon its entropy.

From the above equation it can be seen that free energy changes in any chemical system are inversely related to changes in entropy. A system with high entropy has little free energy and thus little organization. Conversely, a system with low entropy has a greater degree of free energy and thus a greater degree of organization. Therefore, a system with a high degree of organization is capable of performing more work than a system with a low degree of organization. Recall that free energy is generally equated with the ability to do work. A highly organized system has the capability of showing a greater change in free energy than one which is less organized.

The ΔH in the above equation refers to the change in heat during a chemical reaction. The greater the amount of heat given off, the greater the free energy change. A large free energy change indicates that a greater amount of potential energy is released. Therefore, there would be less free energy in the system after reaction than before. Correspondingly, the entropy of the system after reaction would be greater than before.

Consideration of chemical or biological processes in light of the Second Law of Thermodynamics emphasizes two general points. First, for ordinary chemical reactions to proceed, there must be some driving force, whether it is an energy change from a higher to a lower state, or from lesser to greater entropy. This driving force may be defined as the difference in energy levels between two parts of a system. The difference in energy level between glucose and its end products carbon dioxide and water is sufficient to keep the biochemical reactions of a cell going.

Second, all processes in the universe run toward an increase in entropy. This means that the universe as a whole is moving toward greater disorganization. Like a giant clock, wound up some time in the past, the universe is gradually running down. The heat which is given off from the sun and all other stars passes off into space and is lost. The energy is not destroyed. It is simply reduced to a nonusable form. As entropy of the universe increases, free energy decreases. The total amount of usable energy in the universe thus decreases with passage of time.

In the past, various writers have claimed that living organisms defy the Second Law of Thermodynamics. Since living systems grow, reproduce, and metabolize, it has been said that they actually show an increase in free energy and a decrease in entropy. This is true, of course, if the living organism is considered as an isolated system. During periods of growth, for example, an organism builds complex molecules, increases the number of its cells, and shows specialization of certain tissues. The organism, indeed, becomes more highly organized. But no organism is independent of outside energy sources. The energy for all life on earth ultimately comes from the sun. Thus, the sun, green plants, and animals must all be considered as parts of a single system if we are to make any meaningful statements about the thermodynamics of life. Living organisms show increases in free energy only because other parts of the universe show a decrease in free energy.

The balance sheet of the earth-sun system shows that the free energy of the system as a whole is decreasing at an enormous rate. It is only a very small fraction of this energy which is actually captured by living organisms and used to maintain their high levels of organization. Thus, despite activities of living organisms, the Second Law of Thermodynamics still applies to all processes, living or nonliving, in the known physical world.

THE PERFORMANCE OF A CLINICAL ASSAY*

An assay is a test that measures the amount of a substance in a mixture. For example, a physician may want to know how much glucose is present in a patient's blood; how much cholesterol is present; or how much of a specific enzyme (organic catalyst that speeds up chemical reactions) is present.

For each of these cases a specific test has been constructed to determine the amount of the substance present. A successful test will consist of the following characteristics: (1) a chemical reaction that occurs *only* upon the involvement of the substance to be tested; (2) the degree to which the reaction occurs is dependent

* *Source:* Hawk, Philip B., *Physiological Chemistry* (Bernard L. Oser, ed.), (Fourteenth ed.), McGraw-Hill, New York, 1965, pp. 1051–1056; graph from page 1053 (Fig. 29–26).

upon the *amount* of the substance present in the mixture; and (3) the system of measuring the degree of the reaction by instrumentation is *consistent*. Thus we can say that such a test is specific, quantitative, and reproducible.

One system for measuring blood glucose uses a light blue, copper-containing solution that when heated in the presence of glucose turns deep blue. The color change here is dependent upon reaction with glucose. The more glucose, the more reaction, and the more blue color present. The amount of blue can be measured in a spectrophotometer (an instrument that measures the absorption of different wavelengths of light) and is indicative of the quantity of glucose present.

Preparation and performance of the assay. Several test tubes containing the same known volume of the light blue glucose-detecting solution are prepared. An equal volume of each of the following is added next: Water is added to one tube (known as the "reagent blank" because it contains everything except glucose), increasing concentrations of glucose are added to each of the next tubes (these comprise the "standards" because a known amount of standard glucose solution is used), and deproteinized plasma or serum from the patient's blood is added to the next tube (Fig. 5–13).

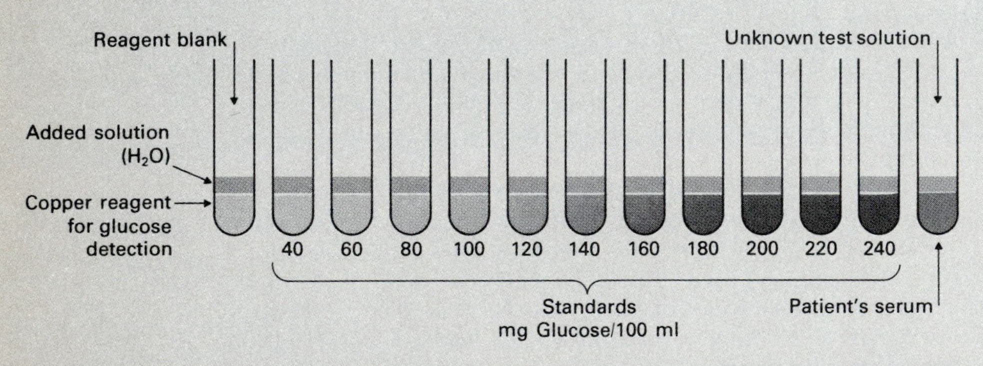

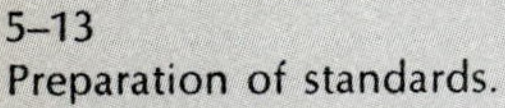
5–13
Preparation of standards.

All tubes are heated for exactly the same amount of time at the same temperature —this is most easily done in a hot water bath. The optical density (amount of light absorbed per unit volume) of the solution in each test tube after allowing the reaction to proceed for the given amount of time is determined in a spectrophotometer, and an "absorption curve" plotting optical density *vs.* wavelength for each of the standard tubes is drawn. Two examples are shown in Fig. 5–14.

The wavelength at which comparisons are made between unknown and standard solutions is 420 mμ. Since the optical density of the solution containing the patient's plasma or serum is known at 420 mμ, the concentration of glucose can be determined from the absorption curves of the standards. For example, an optical density of 0.29 at a wavelength of 420 mμ can be interpreted as a blood glucose level of 100 mg/100 ml.

Note that the precision and accuracy of this clinical test is dependent upon keeping everything constant with the exception of one variable. All volumes must be kept constant, and the time and temperature of the reaction must also be kept constant. The only variable in each test tube is the concentration of glucose.

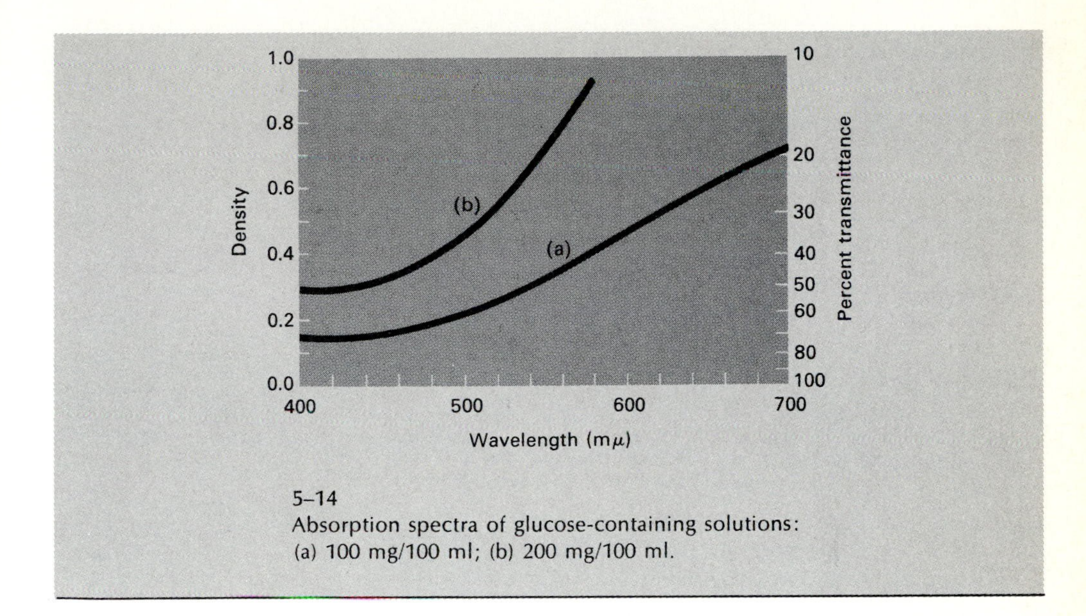

5–14
Absorption spectra of glucose-containing solutions: (a) 100 mg/100 ml; (b) 200 mg/100 ml.

Chapter 6 Acids, Bases, and Neutralization

6–1 INTRODUCTION

The concept of acids and bases plays an essential role in any discussion of the chemistry of living organisms. A great variety of acidic and basic substances occur as intermediate and as end products in biochemical reactions. Amino acids and fatty acids, components of protein and fat respectively, are common acidic substances found in living things. Likewise, purines and pyrimidines, components of nucleic acids, are examples of organic bases. A discussion of acid-base reactions at this point will form a foundation for topics to be considered in later chapters.

6–2 WHAT ARE ACIDS AND BASES?

Acids can be distinguished in several ways. In solution, they show certain characteristics. The sour taste of citrus fruits is due to a high content of citric acid. Sour milk is the result of the accumulation of lactic acid produced by bacteria. Acids turn litmus, an organic dye, from blue to red. Some acids react with metals such as zinc to release hydrogen. Finally, acids neutralize bases to produce water and a salt.

Bases also have certain identifiable characteristics. In solution, they (1) produce a bitter taste; (2) turn red litmus blue; (3) feel soapy when rubbed between the fingers; and (4) neutralize acids. Lye (sodium hydroxide) and milk of magnesia (magnesium hydroxide) are examples of bases.

The above characteristics are useful in identifying acids and bases. However, they say nothing about their chemical nature or the way in which they react. The terms "acid" and "base" were in use long before any satisfactory theory was proposed which explained *why* these compounds possess their particular characteristics.

An acid is any substance which can donate a proton. This property of acids may be represented by the generalized ionization equation below:

$$HA \rightarrow H^+ + A^- \qquad (6\text{–}1)$$

where HA stands for any acid, H^+ represents the proton which the acid is capable of releasing, and A^- symbolizes the negative ion or radical to which the proton is bound in the non-ionized acid molecule. The term *dissociation* is frequently used to refer to the ionization of acids or bases in solution. Thus, Eq. (6–1) could be referred to as a "dissociation" as well as "ionization" equation.

Especially common as proton donors in living systems are *carboxyl* groups, COOH, found in many organic molecules. The carboxyl group dissociates to yield a proton and a negatively charged COO^- ion:

$$-C(=O)OH \rightarrow H^+ + -C(=O)O^- \qquad (6\text{–}2)$$

Carboxyl groups are found as functional units in amino acids and fatty acids, where they give these compounds their acid characteristics. They also occur in a number of other biologically important molecules. Ionization equations for several representative acids are given in Table 6–1.

TABLE 6–1 Ionization equations for representative acids

Name of acid	Ionization equation	Name of (—) charged ion
Hydrochloric	$HCl \rightleftharpoons H^+ + Cl^-$	Chloride ion
Acetic	$CH_3COOH \rightleftharpoons H^+ + CH_3COO^-$	Acetate ion
Phosphoric	$H_3PO_4 \rightleftharpoons H^+ + H_2PO_4^-$	Dihydrogen phosphate ion
Carbonic	$H_2CO_3 \rightleftharpoons H^+ + HCO_3^-$	Bicarbonate ion

A base is any substance that accepts protons. Bases are thus the opposites of acids in their chemical properties. For example, each of the reactions shown in Table 6–1 is to some extent reversible. The negative ionization products of these reactions are capable of accepting protons. They are thus regarded as bases. The generalized equation for a base reaction is:

$$\underset{\text{(Base)}}{A^-} + \underset{\text{(Proton)}}{H^+} \rightarrow \underset{\text{(Acid product)}}{HA} \qquad (6\text{–}3)$$

Some of the most common bases in inorganic chemistry are the *metallic hydroxides*. Upon ionization in water, these substances release positively charged metallic ions and negatively charged hydroxide radicals:

$$\underset{\text{(Sodium hydroxide)}}{NaOH} \rightarrow \underset{\text{(Sodium ion)}}{Na^+} + \underset{\text{(Hydroxide ion)}}{OH^-} \qquad (6\text{–}4)$$

$$\underset{\text{(Magnesium hydroxide)}}{Mg(OH)_2} \rightarrow \underset{\text{(Magnesium ion)}}{Mg^{2+}} + \underset{\text{(Hydroxide ion)}}{2OH^-} \qquad (6\text{–}5)$$

Hydroxide ions are particularly effective bases because they have a very strong attraction for protons. The ions not only show a net electric charge of minus one, but also have three unshared pairs of electrons. These unshared electrons provide areas in the hydroxide ion where protons can be accepted. As a result, hydroxide ions readily pick up any proton which may be in solution (Fig. 6–1). However, many compounds which do not contain hydroxide groups can nevertheless accept protons. Several examples of base reactions are given in Table 6–2.

TABLE 6–2 Representative base reactions

Substance acting as base	Reaction	Name of product
Hydroxide ion	$OH^- + H^+ \rightleftharpoons H_2O$	Water
Acetate ion	$CH_3COO^- + H^+ \rightleftharpoons CH_3COOH$	Acetic acid
Chloride ion	$Cl^- + H^+ \rightleftharpoons HCl$	Hydrochloric acid
Ammonia	$NH_3 + H^+ \rightleftharpoons NH_4{}^+$	Ammonium ion
Water	$H_2O + H^+ \rightleftharpoons H_3O^+$	Hydronium ion

Note that when a base accepts a proton the resulting molecule becomes a potential proton donor. Under the proper circumstances, this newly formed molecule can donate the proton back to some other substance. In other words, it is an acid. As a general rule, then, when a base accepts a proton, it produces a molecule which can act as an acid. All of the products given in Table 6–2 are acids. At first glance, this may seem confusing, since neither water nor the ammonium ion are generally thought of as acidic substances. Further-

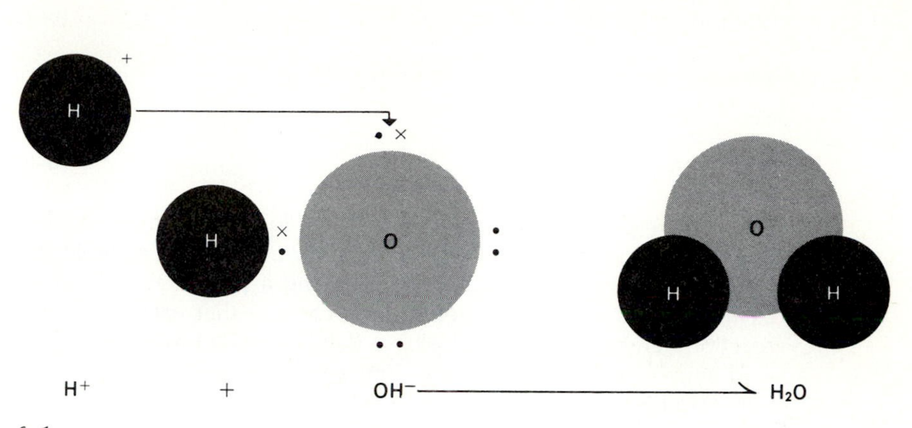

6–1
Electron diagram of the hydroxide ion (OH^-) showing how it attracts protons in solution. The combination of a proton and a hydroxide ion yields a water molecule.

more, in this table water appears in both columns! However, remember that the sole criterion of whether a substance is called "acid" or "base" depends upon its ability to give up or to accept protons, respectively. Many substances can act as acids in one reaction and as bases in another; such compounds are said to be *amphoteric*.

6–3 WATER AND THE STRENGTH OF ACIDS AND BASES

We have seen that the ionization of acids yields a free proton into solution. Yet, experimental evidence indicates that even in solutions of strong acids there are virtually no free protons.

How can this seeming conflict be resolved? Water molecules themselves have a tendency to accept protons. This is because water is a polar molecule with two unshared pairs of electrons. These unshared electron pairs attract free protons:

Unshared electron pairs

$$H^+ \;+\; \longrightarrow \; \begin{matrix} & \circ\circ & \\ \circ\circ & O & \circ\times H \\ & \times\circ & \\ & H & \end{matrix} \;\longrightarrow\; \left[\begin{matrix} & \circ\circ & \\ H\;\circ\circ & O & \circ\times H \\ & \times\circ & \\ & H & \end{matrix} \right]^+ \qquad (6\text{–}6)$$

The addition of an extra proton to a water molecule forms the positively charged *hydronium ion* (H_3O^+). In light of this reaction, the ionization of acids in water is more correctly represented as:

$$HCl + H_2O \rightleftharpoons H_3O^+ + Cl^- \quad (6\text{–}7)$$

Similarly, the accepting of a proton by a base is more correctly shown as:

$$H_3O^+ + B^- \rightarrow HB + H_2O \quad (6\text{–}8)$$

where B^- represents any negative ion. In Eq. (6–7), water acts as a *base* by accepting protons from hydrochloric acid. In Eq. (6–8), the hydronium ion acts as an *acid* by donating a proton to B^-. If we examine these equations carefully, we can see that each involves a reaction between an acid and a base (left-hand side) to yield another acid and base (right-hand side). The general pattern of these reactions can be written as:

$$Acid_1 + Base_1 \rightarrow Acid_2 + Base_2 \quad (6\text{–}9)$$

The extent to which any such reaction occurs depends upon the relative strengths of the acids and bases involved. The strength of an acid is determined by how readily it gives up protons, i.e., how readily the molecule ionizes. The strength of a base is determined by how readily it accepts protons.

As a model, let us consider two acid-base reactions. Note particularly the role of water as a base in each of these reactions.

In water, the ionization of hydrochloric acid is almost complete, and it is a strong acid. In other words, almost 100 percent of the molecules are ionized to H^+ and Cl^-. Thus, virtually all of the protons from the acid have formed hydronium ions, as shown in Eq. (6–7). A molecule of hydrochloric acid is held together by a partially ionic, partially covalent bond involving a shared pair of electrons. However, because of the larger positive charge on its nucleus, the chloride ion has a greater power of attraction for the electron pair. The electron distribution for the molecule is somewhat uneven. There is a greater electron density around the chlorine nucleus than around the hydrogen nucleus. The polar nature of the hydrogen chloride molecule is represented below, where the symbol δ (delta) indicates a partial elec-

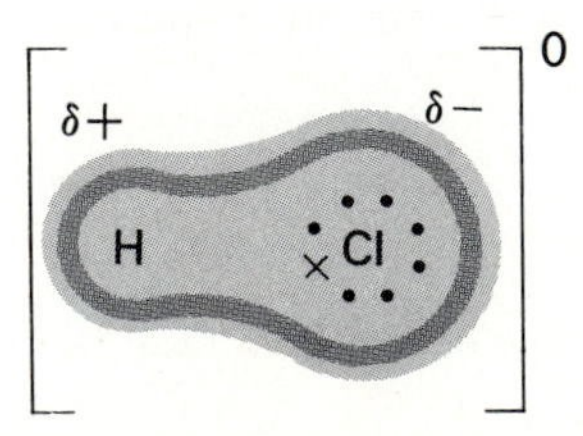

trostatic charge. Note that the molecule as a whole is electrically neutral.

This arrangement provides only a very loose bond between the proton and the chloride ion. The chloride ion therefore releases that proton readily. Only with difficulty can it be made to recombine with it. Since it has such a slight tendency to accept protons, the chloride ion is a weak base.

The complete ionization of hydrochloric acid occurs because water molecules have a much greater attraction for protons than do chloride ions. Water is thus a stronger base than chloride. Molecules of water pull the proton from the acid molecule. In this and other such reactions, the proton is always transferred from the weaker to the stronger base:

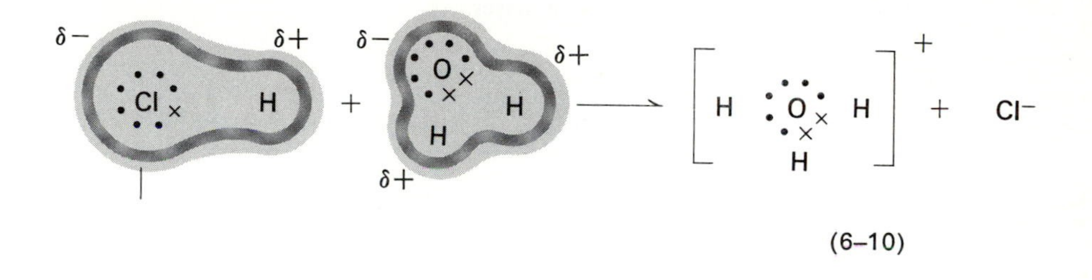

(6–10)

As a second example, consider the ionization of acetic acid. In water, acetic acid ionizes to a considerably lesser extent than hydrochloric acid:

$$HAc + H_2O \rightleftharpoons H_3O^+ + Ac^- \qquad (6\text{–}11)$$

The limited ionization of acetic acid is a result of the great attraction of the acetate ion for protons. Free acetate ions pick up protons readily from any available donors to form the molecule of acetic acid. The water molecules are unable to pull protons from the acetate ions. Thus the acetate ion is said to be a stronger base than water. In the presence of the weaker base water, the acetate ion remains attached to the proton. Large-scale ionization of acetic acid occurs only in the presence of a compound such as ammonia (NH_3), which is itself a stronger base than the acetate ion.

The difference in strength between acetic and hydrochloric acids is thus a difference in the attraction which their respective bases, acetate and chloride ions, have for protons.* Since it ionizes almost

---

* The acetate and chloride ions are often referred to as *conjugate bases*. Any ion which combines with a proton to form an acid is thus the conjugate base of that particular acid.

completely in solution, hydrochloric acid is considered to be a strong acid. On the other hand, acetic acid is considered to be a weak acid since, under similar conditions, it shows only about one percent ionization. For any given concentration, a solution of hydrochloric acid contains many more protons than a solution of acetic acid. This difference is shown diagrammatically in Fig. 6–2. The strength of a given acidic or basic solution can thus be measured in terms of the number of protons released into solution. The stronger the acid, the more protons released; the stronger the base, the fewer protons released. We shall see in Section 6–6 how a measure of the number of protons in solution can provide a scale for determining how acidic or basic a particular solution is.*

TABLE 6–3 Strengths of acids and bases

	Acid		Base		
	Name of acid	Formula	Name of base	Formula	
Increasing strength of acid ↑	Perchloric acid	$HClO_4$	Perchlorate ion	ClO_4^-	Increasing strength of base ↓
	Sulfuric acid	H_2SO_4	Sulfate ion	SO_4^{--}	
	Hydrochloric acid	HCl	Chloride ion	Cl^-	
	Hydronium ion	H_3O^+	Water	H_2O	
	Phosphoric acid	H_3PO_4	Dihydrogen phosphate ion	$H_2PO_4^-$	
	Acetic acid	$HC_2H_3O_2$	Acetate ion	$C_2H_3O_2^-$	
	Carbonic acid	H_2CO_3	Bicarbonate ion	HCO_3^-	
	Ammonium ion	NH_4^+	Ammonia	NH_3	
	Bicarbonate ion	HCO_3^-	Carbonate ion	CO_3^{--}	
	Water	H_2O	Hydroxide ion	OH^-	
	Ethyl alcohol	C_2H_5OH	Ethoxide ion	$C_2H_5O^-$	
	Ammonia	NH_3	Amide ion	NH_2^-	

A chart comparing relative strengths of various acidic and basic substances is given in Table 6–3. Note that the bases fall into an order of increasing strength, just the reverse of the acids. This emphasizes the fact that strength of an acid is inversely related to the strength of the base of which it is composed. In other words, *strong acids exist where protons are bound to weak bases. Weak acids exist where protons are bound to strong bases.*

---

* In using the phrase "protons in solution" we do not mean to contradict the point made previously about the existence of the hydronium ion. Since every hydronium ion contains a proton which it can donate to some other substance, it is customary to disregard the hydronium ion, though its presence is always understood.

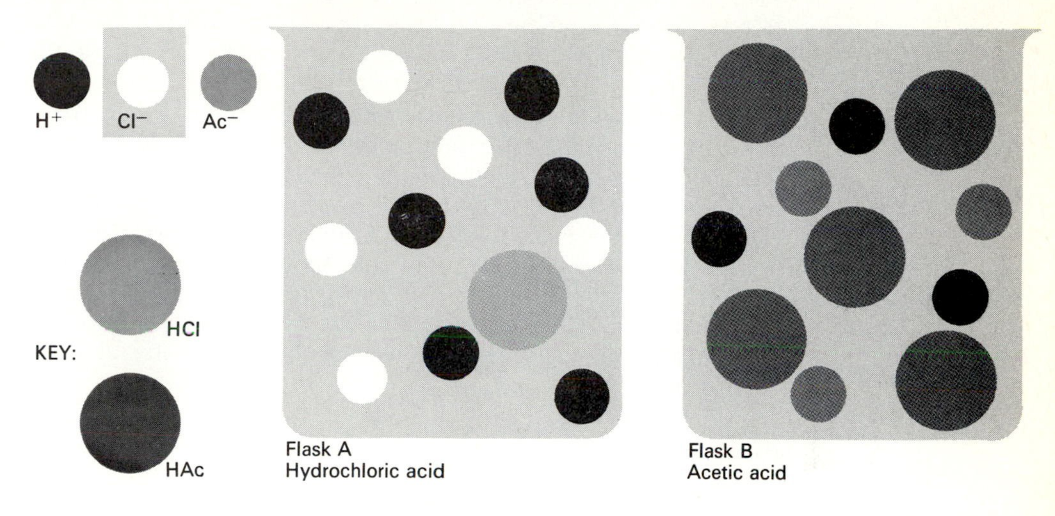

6–2
Diagrammatic representation of the number of protons present in a solution of strong acid (such as hydrochloric) and a solution of weak acid (such as acetic). Because hydrochloric acid ionizes almost 100%, solutions of this acid contain many protons. Acetic acid ionizes only about 1%. As a result, solutions of this acid contain far fewer protons.

In Eq. (6–9), we saw that acid-base reactions yield products which can also act as acids or bases. Considering this principle in relation to the relative strengths of acids and bases, we can now make a generalization. *Acid-base reactions run in the direction of producing a weaker acid and weaker base from a stronger acid and stronger base:*

$$\underset{\text{(Stronger acid)}}{HCl} + \underset{\text{(Stronger base)}}{H_2O} \rightleftharpoons \underset{\text{(Weaker acid)}}{H_3O^+} + \underset{\text{(Weaker base)}}{Cl^-} \qquad (6\text{–}12)$$

Hydrochloric acid is a stronger acid than the hydronium ion. Water is a stronger base than the chloride ion. The arrows show that the equilibrium point of this reaction is shifted heavily to the right. This indicates that the acid and base on the left-hand side of the equation are considerably stronger than the acid and base on the right. This generalization will be particularly important in later discussions of acid-base neutralizations and the action of buffers.

Thus far, we have emphasized the basic nature of water. However, water can also act as an acid by ionizing in the following manner:

$$H_2O + H_2O \rightleftharpoons H_3O^+ + OH^- \qquad (6\text{–}13)$$

Since the hydroxyl ion of each water molecule is a very strong base, the extent to which this ionization occurs in water is very, very slight. The equilibrium for this reaction is shifted so far to the left that only about one molecule out of 554 million undergoes such an ionization. Water is, therefore, a very weak acid. Under most circumstances, it is far easier for a water molecule to accept a proton than to donate one. However, in the presence of a very strong proton acceptor, such as the methide ion (CH_3^-) or the hydride ion (H^-), water can act as an acid. Equation (6–14) shows the reaction between water and the methide ion:

$$CH_3^- + H_2O \rightarrow CH_4 + OH^- \qquad (6\text{–}14)$$

The attraction which the methide ion possesses for a proton is great enough to overcome the strong tendency of hydroxide ions to hold onto protons. In other words, the methide ion is a stronger base than the hydroxide ion of water.

The ammonia molecule has a pair of unshared electrons. Thus, like water, ammonia molecules tend to pick up extra protons. This produces the positively charged ammonium ion $(NH_4)^+$:

$$\begin{matrix} & H & \\ & \circ\times & \\ H\,{}^{\circ}_{\times} & N & {}^{\circ}_{\circ} \\ & \circ\times & \\ & H & \end{matrix} + H^+ \rightarrow \left[\begin{matrix} & H & \\ & \circ\times & \\ H\,{}^{\circ}_{\times} & N & {}^{\circ}_{\circ}\,H \\ & \circ\times & \\ & H & \end{matrix}\right]^+ \qquad (6\text{–}15)$$

Like water, ammonia is a very weak acid. It ionizes to yield a proton and the amide ion very unwillingly. The importance of both water and ammonia in living matter is often their ability to act as bases. The role which these substances play will become apparent in later discussions of acid-base neutralizations and proteins.

6–4 MOLAR AND NORMAL SOLUTIONS

It should be apparent that concentration plays an important role in determining the strength of a given acid or base solution. A concentrated solution of acetic acid, for example, may contain more protons than a dilute solution of hydrochloric acid. Therefore, a few words about two standard types of solutions is in order.

Molar solutions. A molar solution is a means of expressing the quantity of a solute in a given volume of solution. A molar solution of any compound (written as 1 *M*, or *M*) contains one mole of solute in the amount of solvent which produces a total volume (solvent + solute) of one liter. To understand what a molar solution is, we must first define a *mole*.

For any substance whose molecular formula is known, 1 mole contains the Avogadro number of molecules. Avogadro's number is equal to 6.023×10^{23} particles (atoms, molecules, or ions). A *mole*, then, *is a measure of the number of particles in a solution.* For example, if a chemical reaction occurs in which one molecule of a substance reacts with one molecule of another, it is correct to state that *one mole* of one substance reacts completely with *one mole* of the other. Since a mole of each compound contains the same number of molecules, this means that every molecule of both reactants is used up in the course of the chemical reaction between the two compounds.

Referring back to Eq. (6–12), we can summarize this reaction by saying that one molecule of hydrochloric acid donates its proton to one molecule of water to form one hydronium ion and one chloride ion. Or, one *mole* of hydrochloric acid donates its protons to one *mole* of water to form one *mole* of hydronium ions and one *mole* of chloride ions. Since chemical reactions occur on a molecular level, the concept of the mole provides a means of thinking of chemical reactions quantitatively.

TABLE 6–4

Substance	Gram-molecular mass	Number of particles
HCl (gas)	35	6.023×10^{23} molecules of HCl
HCl (in solution)	35	6.023×10^{23} hydrogen ions (protons) and 6.023×10^{23} chloride ions
H^+	1	6.023×10^{23} hydrogen ions (protons)
H_2	2	6.023×10^{23} hydrogen molecules
Glucose ($C_6H_{12}O_6$)	180	6.023×10^{23} glucose molecules

How much of any substance constitutes a mole? In other words, how much of any substance contains 6.023×10^{23} particles? The answer to this question turns out to be easier than might first be imagined. One *gram-molecular mass* (the molecular mass expressed in grams) of all compounds contains 6.023×10^{23} molecules. One mole of any compound is thus equal to the gram-molecular mass (Table 6–4).

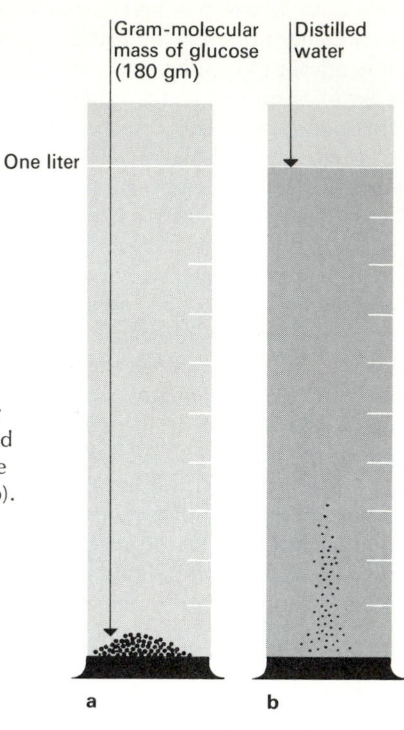

6–3
Preparation of a molar solution. A gram-molecular mass of the solute is placed in an empty graduated cylinder (a). Enough water is then poured into the cylinder to bring the total volume up to one liter (b).

Moles are most useful in expressing concentrations of various solutions. For example, if one mole of glucose ($C_6H_{12}O_6$, gram-molecular mass, 180 gm) is dissolved in water and the total volume of the solution (solvent + solute) brought to one liter, a one molar solution of glucose is obtained (see Fig. 6–3). Similarly, a one molar solution of the salt magnesium sulphate would contain 120.39 gm of solute in a total volume of one liter.

To prepare a molar solution, a number of grams of solute equal to the molecular mass of the solute is placed into a measuring container. Enough water is added to make a total volume of one liter. Suppose, for example, we wish to prepare a one molar solution of glucose. The gram-molecular mass of glucose is 180 gm. We place 180 gm of the compound in a graduated cylinder (Fig. 6–3a). By adding enough water to bring the total level in the cylinder up to the one liter mark, a 1*M* solution of glucose is obtained.

Solutions can also be prepared in concentrations greater or less than one molar. For example, if one-tenth of a mole of glucose (18.0 gm) is dissolved in water and the volume adjusted to one liter, the solution is referred to as 0.1 molar (0.1 *M*). A 2 molar (2 *M*) solution of glucose can be prepared by dissolving 360 (2 × 180) gm of the compound in distilled water and adjusting the volume to one liter.

A simple example will illustrate the advantage of knowing the exact number of particles of solute in a solution. Suppose that a biologist wants to know how efficient a certain catalyst is in a reaction. What he really wants to know is the number of molecules which the catalyst can act upon in a given unit of time. By using solutions of known molarity, it is easy to make such a calculation. Suppose he adds a 0.1 *M* solution of catalyst to a 1 *M* solution of reactant. Stopping the reaction after a certain period of time, he can measure the change in concentration of the reactant. Since he is using molarity to express concentration, he knows immediately how many particles of reactant have been used up in the reaction. Knowing also the concentration of catalyst in molarity, it is a simple step to calculate the number of reactant molecules acted upon per molecule of catalyst. This, in turn, measures the efficiency of the catalyst.

Normal solutions. A normal solution contains one *gram-equivalent* of the solute in one liter of solution. One gram-equivalent of a substance is the mass which (1) *as an acid* contains one gram-atom of replaceable hydrogen, that is, hydrogen which can be dissociated from the molecule or "replaced" in the molecule by some other positive ion; (2) *as a base* reacts with one gram-atom of hydrogen; or (3) *as a salt* is the product of a reaction involving one gram-atom of acid hydrogen. Since normal solutions are most frequently used in acid-base reactions, we shall confine our present discussion of gram-equivalents to acids and bases.

Suppose that we wish to prepare a one normal (1 *N*) solution of hydrochloric acid. Its gram-molecular mass is 36.5 gm. Of this, one gram is replaceable hydrogen, since each molecule contains one hydrogen atom. We divide the number of gram-equivalents of hydrogen into the total molecular mass. This gives us the number of grams which should be dissolved in a liter of solution. In this case the number is equal to the gram-molecular mass ($36.5/1 = 36.5$ gm). Thus, for HCl, a 1 *N* solution is the same concentration as a 1 *M* solution.

On the other hand, if we wished to prepare a 1 *N* solution of sulfuric acid, we would find a slightly different situation. The gram-molecular mass of H_2SO_4 is 98.082 gm. Each molecule of this acid, however, contains two gram-equivalents of hydrogen. Hence, to prepare a 1 *N* solution we would have to divide 98.082 gm by 2, which gives 49.041 gm/liter.

Normality of the various metallic hydroxides is calculated by dividing the total gram-molecular mass of a given substance by the number of hydroxide ions which are capable of reacting with hydrogen. Thus, a normal solution of sodium hydroxide [NaOH] would contain 40/1 or 40 gm of solute in one liter of water.

In working with acids and bases, normal solutions have the advantage over molar solutions in that, measured in normals, a strong acid will react completely with an equal amount of a strong base.

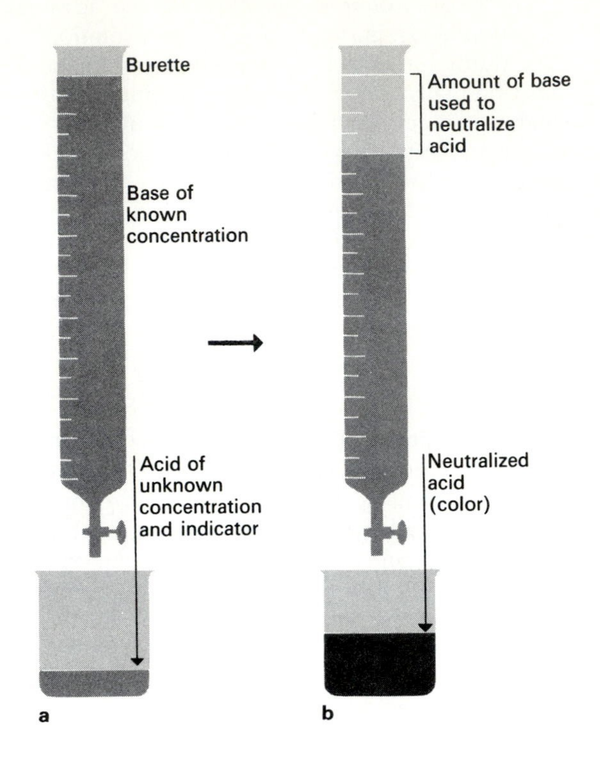

6–4
The concentration of an acid can be determined by titration. A base of known concentration is slowly added to an acid in the beaker. The acid contains an *indicator,* a compound whose color (or lack of it) indicates whether the solution is acidic, basic, or neutral. When enough base has been added to neutralize the acid, the indicator changes, turns color, showing that the solution in the beaker has passed from acid through neutrality to basic. By recording how much base was added, the concentration of the acid can be estimated. Most indicators are organic compounds which are sensitive to changes in hydronium ion concentration.

The reason for this can be seen by examining the following equations:

$$H_2SO_4 + 2NaOH \rightarrow Na_2SO_4 + 2H_2O \qquad (6\text{–}16)$$

$$HCl + NaOH \rightarrow NaCl + H_2O \qquad (6\text{–}17)$$

Equation (6–16) indicates that one mole of sulfuric acid reacts with two moles of sodium hydroxide to produce one mole of sodium sulfate and two moles of water. If the solutions of sulfuric acid and sodium hydroxide were of the same molarity, twice the volume of sodium hydroxide would be required to react with all the hydrogen and sulfate ions. Equation (6–17), on the other hand, shows that one mole of hydrochloric acid will react completely with one mole of sodium hydroxide. In one case, complete reaction results from a one-to-one ratio (by volume) of acid to base. In another, a one-to-two ratio is needed. Work in biochemistry often calls for reactions where all of the acid and base involved must be used up, so that no protons are left in solution. Using molar solutions, the volume of acid or base needed for every reaction must be calculated. This work is eliminated by employing solutions of known normality.

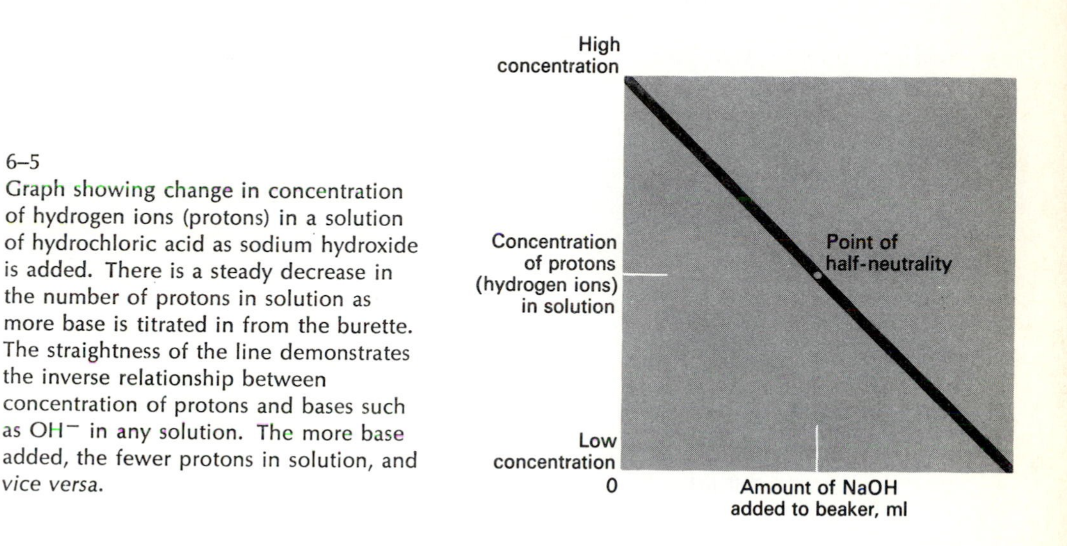

6–5
Graph showing change in concentration of hydrogen ions (protons) in a solution of hydrochloric acid as sodium hydroxide is added. There is a steady decrease in the number of protons in solution as more base is titrated in from the burette. The straightness of the line demonstrates the inverse relationship between concentration of protons and bases such as OH^- in any solution. The more base added, the fewer protons in solution, and *vice versa*.

6–5
SALTS AND ACID-BASE NEUTRALIZATION

Neutralization reactions decrease the total number of molecules of strong acid or base in solution. Any reaction where a stronger acid and base yield a weaker acid and base is a neutralization reaction. The weaker the acid and base produced by such a reaction, the more complete the neutralization (see Fig. 6–4).

The reaction of a strong acid (hydrochloric) and a strong base (sodium hydroxide) illustrates the principle involved in neutralization. Such a reaction produces a salt and water:

$$HCl + H_2O \rightarrow H_3O^+ + Cl^- \qquad (6\text{–}18)$$

$$NaOH \rightarrow Na^+ + OH^- \qquad (6\text{–}19)$$

Because hydroxide ions act as a strong base, they remove the extra proton from hydronium. Thus, we get a reaction which is central to

TABLE 6–5 Acid-base neutralizations

Reaction	Salt formed	Water	Name of salt
$2NaOH + H_2SO_4 \rightarrow$	Na_2SO_4	$+ 2H_2O$	Sodium sulphate
$KOH + HCl \rightarrow$	KCl	$+ H_2O$	Potassium chloride
$2KOH + H_2SO_4 \rightarrow$	K_2SO_4	$+ 2H_2O$	Potassium sulphate
$Ca(OH)_2 + 2HCl \rightarrow$	$CaCl_2$	$+ 2H_2O$	Calcium chloride
$Ca(OH)_2 + H_2CO_3 \rightarrow$	$CaCO_3$	$+ 2H_2O$	Calcium carbonate

acid-base neutralization:

$$H_3O^+ + OH^- \rightarrow H_2O + H_2O \qquad (6\text{–}20)$$

This leaves free sodium and chloride ions in solution. Since water molecules ionize so slightly, the protons donated by the hydrochloric acid and the hydroxide ions from the sodium hydroxide are "trapped." If the water is evaporated, crystals of the salt sodium chloride are obtained (see also Table 6–5).

6–6 THE pH SCALE

So far, our discussion of acids and bases has been in general, qualitative terms. In scientific work, however, it has been found helpful to devise a scale to measure how acidic or basic a given solution is. This, the *pH scale,* is based upon the concentration of hydrogen ions* in a liter of solution.

The pH scale runs from 0 to 14. The lower numbers refer to acid solutions. The higher numbers refer to basic solutions. The midpoint in the scale is 7, the pH of water. At this point the concentration of hydrogen ions equals the concentration of hydroxide ions. Any solution with a pH of less than 7 has more hydrogen than hydroxide ions in solution. Conversely, any solution with a pH of more than 7 has fewer hydrogen than hydroxide ions in solution.

---

* pH has traditionally been defined in terms of the "concentration of hydrogen ions" per liter of solution. Since hydrogen ions are protons, pH could just as well be discussed in terms of "concentration of protons" per liter of solution. To be in line with other descriptions of pH found in the literature, the term "hydrogen ion" will be used throughout this section. The pH scale, of course, actually measures the concentration of hydronium ions (see Section 6–3, page 93).

The pH scale is based on actual calculations of the number of hydrogen ions in solution. This can be experimentally determined in several ways. One of the most convenient means is by the use of indicators. The figures obtained in this way can then be converted to a value on the pH scale. The pH scale is thus a yardstick for measuring the number of hydrogen ions in solution, since the concentration of hydrogen ions in solution is expressed as moles of hydrogen ion per liter.

Why was the number 7, the pH of water, chosen as the mid-point on the pH scale? Careful measurements show that a liter of pure water contains one ten-millionth of a mole of hydrogen ions, or $1/10{,}000{,}000 \times 6.023 \times 10^{23}$ particles of H^+. It is quite awkward to write all of the zeros involved in such numbers as 1/10,000,000. It is much easier to use an exponential system and write this number as 10^{-7}. To arrive at a pH value, the negative exponent (−7) is converted into the positive number 7. Thus, water has a pH of 7. Similarly, hydrogen ion concentrations of 1/10,000 (10^{-4}) moles per liter and 1/100,000,000 (10^{-8}) moles per liter are written as pH 4 and pH 8 respectively. Since the number is actually a negative exponent, *the smaller the exponential value, the greater the concentration of hydrogen ions.* A diagram of the pH scale, showing both actual concentrations and corresponding pH values, is given in Fig. 6–6.

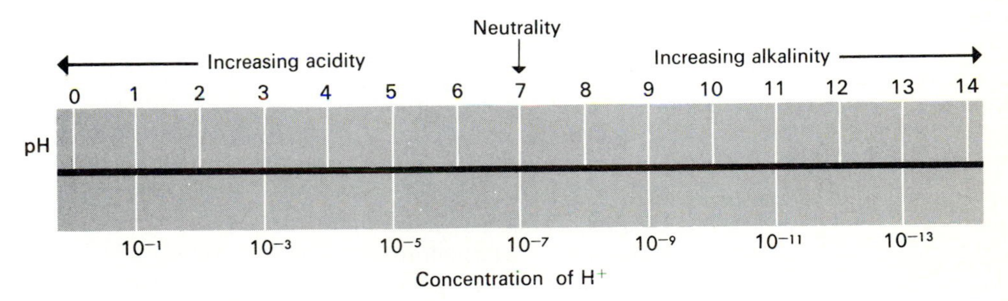

6–6
Diagrammatic representation of the pH scale. The pH value is at the top. At the bottom is the actual concentration of hydrogen ions expressed in moles per liter.

The pH scale gives values as high as one-tenth (10^{-1}) of a mole of hydrogen ions per liter, to values as low as one hundred-trillionth (10^{-14}) moles per liter. The pH of most acids and bases falls somewhere between this span.

The pH scale is a *logarithmic progression.* This means that numerical values on the scale are not like an arithmetical progression,

TABLE 6–6 Hydrogen and hydroxyl ion concentration in a solution of HCl

Concentration of HCl, moles/liter	Concentration of OH ions	Concentration of hydrogen ions from dissociation of H_3O^+*	Point on pH scale
0.1	10^{-13}	10^{-1}	1
0.01	10^{-12}	10^{-2}	2
0.001	10^{-11}	10^{-3}	3
0.0001	10^{-10}	10^{-4}	4
0.00001	10^{-9}	10^{-5}	5
0.000001	10^{-8}	10^{-6}	6
0.0000001	10^{-7}	10^{-7}	7

* Note that the pH scale is determined as the concentration of hydrogen ions, not as the concentration of hydroxyl ions. However, in an aqueous solution the concentration of hydrogen ions and hydroxyl ions always equals 10^{-14} moles/liter, by definition. Thus, as one increases, the other decreases by the same amount, as shown by comparing columns two and three.

where the value of 2 is twice that of one, or a value of 3, three times that of one. Rather, numbers on the pH scale are based on powers of ten. Logarithmic scales are quite useful when large changes in quantity must be measured. On the pH scale, a pH of 2 indicates ten times fewer hydrogen ions than a pH of 1, rather than half as many. A pH of 3 indicates ten times fewer hydrogen ions than a pH of 2, and a hundred times fewer hydrogen ions than a pH of 1. A pH of 4 indicates ten times fewer hydrogen ions than a pH of 3, a hundred times fewer than a pH of 2, and a thousand times fewer than a pH of 1. Thus the number of hydrogen ions released by compounds of successively higher pH decreases by a factor of ten with each step up the scale. This can be illustrated by referring to Table 6–6.

A 0.1 *M* solution of hydrochloric acid would have a pH of 1, and a 0.01 *M* solution a pH of 2. 0.1 *M* means one-tenth of a mole of hydrogen ions (i.e., hydronium ions) per liter, while 0.01 *M* means one-hundredth of a mole per liter. Thus, 0.01 represents a tenfold *decrease* in hydrogen ion concentration over 0.1, and is represented by the difference between the expressions 10^{-1} and 10^{-2}. Similarly, an HCl solution of 0.000001 *M*/liter (10^{-6}) would represent a millionfold dilution of a solution at 0.1 *M*/liter concentration. Each step up the pH scale thus represents a tenfold decrease in the hydronium ion concentration, and each step down a tenfold increase. A similar table for basic solutions is shown in Table 6–7.

Tables 6–6 and 6–7, respectively, show that an acid solution cannot have a hydrogen ion concentration of less than 10^{-7} *M*/liter, nor can a basic solution have an hydroxyl ion concentration of less than 10^{-7} *M*/liter. In a way this is simply a restatement of the pH scale.

TABLE 6–7 Hydrogen and hydroxyl ion concentrations in a solution of NaOH

Concentration of NaOH, moles/liter	Concentration of OH ions	Concentration of hydrogen ions from dissociation of H_3O^+*	Point on pH scale
0.0000001	10^{-7}	10^{-7}	7
0.000001	10^{-6}	10^{-8}	8
0.00001	10^{-5}	10^{-9}	9
0.0001	10^{-4}	10^{-10}	10
0.001	10^{-3}	10^{-11}	11
0.01	10^{-2}	10^{-12}	12
0.1	10^{-1}	10^{-13}	13
0.0	10^{0}	10^{-14}	14

* See note for Table 6–6.

But there is something else of importance to see here, too. With a solution of HCl and pure water, the pH could never rise above 7 nor could the pH fall below 7 with a solution of NaOH and pure water. The reason for this is simple. For the hydronium ion concentration to be less than that of pure water (10^{-7}), some proton acceptor must be present to pull the H^+ away from the H_3O^+. Sodium hydroxide and other bases serve this function. Thus, for a solution of HCl to have fewer H_3O^+ ions in it than pure water, some base must be present. Any acid solution, no matter how dilute, must have more hydronium ions than pure water. If it does not, it is not an acid, by definition. It would be impossible to have an aqueous solution of hydrochloric acid which had less hydrogen ions than 10^{-7} *M*/liter. Similarly, no solution of pure NaOH and water could have a hydronium ion concentration of more than 10^{-7}. In order for that to occur, acid would have to be added, providing a supply of hydrogen ions which could combine with H_2O to yield hydronium ions. Were any additional molecules of base (for instance, OH^-) around, they would absorb the H^+ and thus keep the pH of the solution at 10^{-7} or higher.

The following statements summarize how pH scale values relate to the actual strength or weakness of acidic and basic solutions:

1. The pH scale ranges from 0 to 14. Solutions having a pH of less than 7 are acidic, those with a pH greater than 7 are basic.
2. The midpoint of the scale, pH 7, represents neutrality. Here, the number of hydrogen ions equals the number of hydroxide ions. The pH of water is 7.
3. The pH scale is a logarithmic one. A change of one unit in pH corresponds to a tenfold change in hydrogen ion concentration.

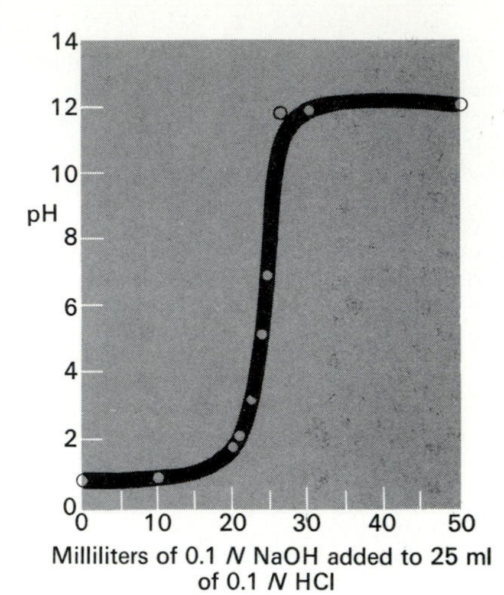

6–7
Titration curve for 0.1 *N* NaOH added to 25 ml of 0.1 *N* HCl. In this reaction, change in pH of the solution is more pronounced than in a weak acid system such as that shown in Fig. 6–8.

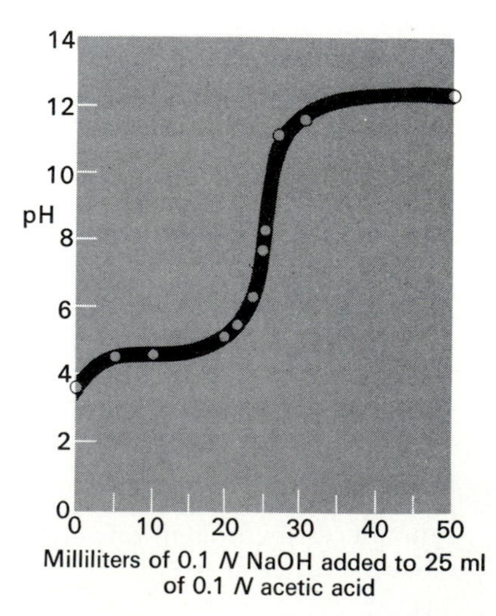

6–8
Titration curve for 0.1 *N* NaOH added to 25 ml of 0.1 *N* HAc. The addition of up to 20 ml of base does not change concentration of hydrogen ions nearly so much as addition of a similar volume of base to a strong acid such as HCl. Acetic acid—sodium hydroxide represent a buffer system.

4. The pH scale is a standardized means of expressing the acidity or alkalinity of any solution. It provides a frame of reference by which the concentration of hydrogen ions in various solutions can be judged.

It is quite possible to have a pH outside of the normally encountered range of 1 to 14. However, these extremes rarely occur in biochemical work. They are certainly not found in living organisms. As a matter of fact, the large majority of pH measurements within living organisms lie well between 6 and 8. It is true that the stomach, with large quantities of hydrochloric acid, has a pH of 1 or 2, and that certain bacteria require a very acid medium in which to live. However, these are exceptions to the rule. The majority of plants and animals are restricted to an internal environment which varies only slightly to one side or the other of neutrality.

6–7 BUFFER SYSTEMS

If we graph the changes in pH of a strong acid solution as a base is added to it, a curve is obtained like that shown in Fig. 6–7. Note that the addition of 10 ml of sodium hydroxide does not markedly change the pH. Addition of 20 ml, however, produces a noticeable upward swing in the curve. This indicates a substantial decrease in concentration of hydrogen ions in solution. Addition of a total of 25 ml raises the pH to seven, while further addition of base produces a dramatic increase in pH. The change in shape of the curve at the 25 ml mark gives the impression that the actual concentration of hydrogen ions is decreasing at an uneven rate. However, this is not the case. Recall that a jump from pH 2 to pH 4 does not indicate a change in hydrogen ion concentration equal to a jump from pH 6 to pH 8.

Now, consider Fig. 6–8. Here, sodium hydroxide is added to a solution of acetic acid, a much weaker acid than hydrochloric. The shape of this curve is not unlike that shown in Fig. 6–7. However, there is one important difference. In Fig. 6–7, the graph line begins at a lower pH. By the time 20 ml of sodium hydroxide have been added, the pH has risen from 1 to 2, a substantial reduction in hydrogen ion concentration. On the other hand, the graph line for the weak acid in Fig. 6–8 begins at a higher pH. With the addition of the first 5 ml of base there is a slight increase in pH. However, the addition of the next 15 ml brings about a pH change from only 4 to 5. A change from pH 4 to pH 5 represents considerably less change in actual hydrogen ion concentration than a change from pH 1 to pH 2. Thus, over a certain range, the acetic acid system maintains relatively constant pH, despite the addition of a strong base.

The acetic acid, sodium hydroxide reaction is an example of a *buffer system,* or a *buffered reaction.* Buffer systems resist changes in pH when a strong acid or base is added. A buffer system is often composed of a weak acid and one of its salts, such as acetic acid and sodium acetate.

How does a buffer system operate? Let us consider the reaction of acetic acid and sodium hydroxide as an example. In water solutions, pH depends upon the concentration of hydrogen ions. The addition of acid will increase the number of such ions by the reactions

$$HA \rightarrow H^+ + A^- \qquad (6\text{–}21)$$

Addition of base will decrease the number of hydrogen ions:

$$B^- + H^+ \rightarrow BH \qquad (6\text{–}22)$$

When a weak acid (such as acetic) is placed in water, it ionizes only slightly, releasing relatively few hydrogen ions into solution:

$$HAc \rightleftharpoons H^+ + Ac^- \qquad (6\text{–}23)$$

Here, the non-ionized acetic acid molecules (HAc) far outnumber the H^+ and Ac^- ions. Molecules of acetic acid thus act as reservoirs of hydrogen ions. Water molecules cannot bring about further ionization, since they are a weaker base than acetate.

AN EXAMPLE OF A BIOLOGICAL BUFFER SYSTEM

Blood is maintained at a relatively constant pH of about 7.4. One of the contributing factors is the presence of a high concentration of bicarbonate ion, HCO_3^-, which is important in the following reversible reaction sequence:

$$\underset{\text{Carbon dioxide}}{CO_2} + \underset{\text{Water}}{H_2O} \rightleftharpoons \underset{\text{Carbonic acid}}{H_2CO_3} \rightleftharpoons \underset{\text{Hydrogen ion}}{H^+} + \underset{\text{Bicarbonate ion}}{HCO_3^-}$$

The equilibrium operates in conjunction with the equilibrium of aqueous CO_2 in the blood with gaseous CO_2 in the lungs.

However, hydroxide ions are a stronger base than acetate ions. Addition of sodium hydroxide thus shifts (indirectly) the equilibrium of reaction (6–23) to the right, which results in further ionization of acetic acid. The result is that all of the hydroxide ions added in the form of sodium hydroxide will be combined with protons. They are trapped in the much weaker base, water. This reaction is summarized

$$Na^+ + OH^- + HAc \longrightarrow H_2O + Ac^- + Na^+ \qquad (6\text{–}24)$$

Ions of acetate (Ac^-) and sodium (Na^+) are left in solution. The important feature of this reaction is that the concentration of hydrogen ions in solution (the pH) has remained much the same as it was before the hydroxide ions were added.

What happens when a strong acid such as HCl is added to the sodium acetate, acetic acid system? The salt sodium acetate dissociates to a very large extent. Thus, there are many acetate ions in solution. Since acetate is a strong base, it will combine readily with hydrogen ions. Thus, any hydrogen ions added to the system will immediately be bound by acetate to form acetic acid. Since it does not ionize to any great extent, the acetic acid formed in the way serves as a trap for hydrogen ions. Despite the addition of large numbers of hydrogen ions, the pH of the solution remains relatively constant.

A buffer system provides a built-in mechanism to counteract either an increase or decrease in hydrogen ion concentration. Every buffer system has a certain pH range within which it can operate most efficiently. The acetic acid-sodium acetate system, for example, has

In hyperventilation, which occurs when a person breathes deeply and quickly for an extended period of time, CO_2 is blown off faster than it is produced and the concentration of H_2CO_3 in the blood is lowered; the pH therefore increases (respiratory alkalosis) as the equilibrium shifts to the left.

If, for some reason, excessive CO_2 is present in the blood, as for example, in hypoventilation, the equilibrium will shift to the right and the pH of blood will be lowered (respiratory acidosis) until the CO_2 can be released from the lungs as gas.

In both cases the red blood cells and kidneys function to restore the original equilibrium.

Red blood cells contain the enzyme carbonic anhydrase, a catalyst for the reaction $CO_2 + H_2O \rightleftharpoons H_2CO_3$ which is otherwise a relatively slow reaction. This catalysis is important because it provides a mechanism for quick hydration of CO_2 as it is its greatest effectiveness at about pH 4 to 5. This means that at an initial pH of 4.5, large amounts of acid or base can be added without an appreciable change in total pH of the system. When too much base is added to the solution, as indicated in Fig. 6–8, at about 25 ml most of the acetic acid molecules have ionized. Addition of more base greatly increases the pH, since the reservoir of hydrogen is now exhausted.

An analogy may help to summarize the action of buffers. Consider the springs of an automobile. If they are subjected to a bump, they absorb the shock. The entire system, including the body of the car and its passengers, remains relatively unaffected. Equilibrium has been maintained. Like the springs of a car, buffer systems absorb the "shock" of a sudden, drastic change in hydrogen ion concentration. Occasionally, of course, the car strikes a very large object. The springs are unable to handle such a shock, and the passengers are jolted. Similarly, we can see from Fig. 6–8, that a given buffer system can maintain a balance only within a certain limit. Outside of this limit, the buffer system is ineffective in maintaining a constant pH.

Most life processes depend upon regulation of pH within the individual cell as well as in the medium which surrounds the cell. Living organisms cannot tolerate large fluctuations in pH. In mammals, for example, the proper functioning of hemoglobin molecules in carrying oxygen to the tissues depends upon the maintenance of a blood pH around 7.4. Despite the constant release of acidic and basic

removed from tissues and for quick dehydration of H_2CO_3 to gaseous CO_2 in the lungs where it can be excreted.

The kidneys aid in maintaining the pH of blood at 7.4 by excreting acidic urine and NH_4^+ (ammonium ion) under conditions in which blood pH is decreased. If the blood becomes alkaline (increased pH), the kidneys will counteract this by excreting Na^+ (sodium ions), HCO_3^-, and the dissociated forms of other weak acids.

Sources: White, Abraham, Philip Handler, and Emil L. Smith, *Principles of Biochemistry* (Fifth ed.), McGraw-Hill, New York, 1973, pp. 889–894; and *Biology Today*, CRM Books, 1972, pp. 382, 395.

metabolic* wastes within the organism, the pH of the blood remains surprisingly constant. There are many buffer systems found in the cells and the tissue fluids of all living organisms. By their action to maintain a constant pH, these buffer systems limit the damage that sudden addition of H^+ or OH^- could cause to delicate membranes and cells in the living system.

---

* Metabolism is the sum total of chemical activity occurring within a living organism.

Chapter 7 The Chemical Composition of Living Material

Most of the matter that makes up living organisms is found in the form of molecules or ions rather than as elements. Living systems are made up of aggregations of molecules. These molecules are divided somewhat arbitrarily into *inorganic* and *organic*. These two groups will be considered separately.

7-1 INORGANIC SUBSTANCES IN LIVING MATTER

There is little doubt that *water* is the most prevalent inorganic compound in living organisms. Life is absolutely dependent upon the presence of this versatile substance. It is true that some animals, such as *Dipodomys*, the kangaroo rat, can get along without *drinking* water. However, *Dipodomys* must still extract from its food the water which is necessary for its metabolic processes. It has also evolved elaborate mechanisms to conserve water by reducing the amount of evaporation from the body and by eliminating very little water in body wastes.

There is no definite amount of water that must be present in living matter as long as at least a small amount of it is there. This amount may vary from 20 to 95 percent from one organism to another, or even from one tissue to another. A man weighing 150 lb contains about 100 lb of water.

Water serves several important functions in living organisms. It is the medium in which many substances are dissolved. In solution, many salts are ionized and become chemically reactive. In water large organic molecules can remain suspended and evenly dispersed. Thus they are better able to come into contact with other ions and compounds in order to interact chemically with them. Since water is such an excellent solvent, it is also used as a transporting agent. Nutrients in solution are carried to the cells, and wastes in solution are removed. Urine, for example, is simply water with dissolved excretory compounds concentrated within it.

Water also gives a living organism the ability to resist sudden changes in temperature which might otherwise be fatal. Water absorbs and releases heat very slowly.

If water is slow to change its temperature from cold to hot, it is also slow to do the reverse. Hot water bottles are used to keep beds warm at night. England's climate is far milder than would be expected for a country lying in the same latitude as Southern Alaska and Moscow. The warm currents of the Gulfstream are responsible for the temperate conditions. Although these waters travel thousands of miles from tropical regions to the North Atlantic, they still remain relatively warm. This shows well the high heat capacity of water.

Thus, in living organisms, water not only prevents rapid heating, it also prevents rapid cooling. It thereby aids in keeping a steady body temperature, often vital to the proper chemical functioning of an organism. Even in death, the bodies of larger organisms retain their warmth for hours, due to their high water content.

Water also serves as a lubricant in the body. Not only are the bones, ligaments, and tendons constantly rubbing against each other as they are moved, but most of the internal organs touch and slide over each other. The liver rubs against the diaphragm during breathing, and the pancreas rubs against the stomach. Annoying and perhaps painful friction is avoided by the dampness of these organs. The body also contains thousands of *mucous* glands, which secrete *mucus*. Mucus contains a great deal of water and is the body's most widely used lubricant. It is produced in great quantities in the nasal passages and in the lining of the digestive tract.

The senses of taste and smell are both dependent upon water. If the surfaces of the nose and tongue were perfectly dry, it would be impossible to perform either of these sensory functions. We cannot taste or smell a compound until its molecules have entered at least partially into solution. Then the nerve endings can react chemically to their presence.

External respiration also depends upon the presence of water. If the surfaces of the air sacs (alveoli) of the lungs are not damp, oxygen cannot dissolve. Thus it cannot enter the blood stream by diffusion. On the other hand, a fish can live a long time out of water as long as its gills are kept damp, so that oxygen from the air can be dissolved. In both plants and animals, it is essential that the surfaces of all living cells be kept moist in order that an exchange of materials may occur.

Water is used as a reactant by living organisms in many chemical processes. In digestion, for example, the chemical addition of water molecules to a nutrient such as table sugar (sucrose) in the presence of the enzyme sucrase results in the splitting apart of the sucrose molecule into two smaller ones. Water is used in a similar manner in the digestion of proteins and lipids (fats). Water molecules are also involved in the manufacture, or *synthesis,* of certain large molecules from smaller ones in living organisms. In Chapter 8, the role of water in digestion and synthesis will be discussed in greater detail.

Besides water, all other inorganic compounds in living matter fall under the general heading of acids, bases, or salts (Chapter 6). The most plentiful salts in the human body are those containing phosphorus and calcium, both of which are deposited in bone and teeth. Phosphorus is also a component of the nucleic acids *DNA* and *RNA,* as well as the very important energy source, *adenosine triphosphate,* or *ATP*. Elements such as sulphur, sodium, chlorine, magnesium, and iron occur in much lesser amounts (Table 7–1).

In the case of iron, 4 atoms are found in every molecule of *hemoglobin*. Hemoglobin is the oxygen-carrying protein which gives blood its red color. If this iron is extracted from the body by chemical means, about enough to make 2 small nails is obtained. Do not let this slight amount mislead you. The iron is still extremely important, for without hemoglobin, blood can carry far less oxygen. Even the loss of a slight amount of iron is enough to bring on severe anemia, which results in general weakness.

When the necessary amount of an element in an organism is extremely small, it is called a *trace element*. Iodine is a trace element vital to the functioning of the thyroid gland, which regulates the rate of the body's chemical activity, or *metabolism*. A slight iodine deficiency may result in a huge swelling on the neck in the region of the gland, a condition known as simple goiter. Complete absence of iodine is fatal. Yet the total amount of iodine in a man's body is only about 1/2,500,000 of the entire body weight. As is the case with all minerals, iodine must be supplied to the body in the diet.

It must be clearly understood that these elements are present in the body of a plant or animal only in chemical combination with other elements or as ions. The presence in a living organism of very small amounts of any of the *elements* iodine, chlorine, sodium, or phosphorus would be instantly fatal. However, by virtue of their being in chemical combination with some other element or elements,

TABLE 7–1 Some elements in living organisms

Name of element	Some uses by living organisms
Oxygen	Final hydrogen acceptor in respiration of some forms
Nitrogen	Part of all protein and nucleic acid molecules
Hydrogen	Part of all foods and water
Carbon	Part of all foods; basic atomic building block of all organic compounds
Sulfur	Found in many proteins
Phosphorus	Found in many proteins; especially prominent in nerve tissue, bone, adenosine triphosphate, and nucleic acids
Iron	Essential part of hemoglobin molecule and other important heme molecules such as the respiratory enzymes
Calcium	Structural atom in bone and teeth; essential in blood clotting and for the synthesis of certain hormones
Sodium	Structural atom in bone and tissue; essential in blood in salt to maintain proper osmotic balance; important for nerve conduction
Potassium	Essential for growth; important for nerve conduction
Iodine	Production of thyroid hormone in thyroid gland
Chlorine	Combined with sodium to form NaCl, an important salt

their electron clouds have been rearranged. Thus the chemical properties they possess as elements are changed.

7–2 ORGANIC SUBSTANCES IN LIVING MATTER

Both inorganic and organic substances are essential for life. However, we still tend to think of living material as being primarily *organic,* and so it is.

It was once thought that organic substances were only produced by the chemical processes within living things. However, in 1828, the German chemist Friedrich Wohler transformed ammonium cyanate into the organic compound urea in the laboratory. Today it is known that, of the over 1 million organic compounds, only about 5 percent are produced by the metabolism of living organisms.

All organic compounds contain the element *carbon.* In fact, organic chemistry is often called "the chemistry of carbon." If the compounds produced with all the other elements except carbon are counted, approximately 60,000 can be found. But over 1 million are known with carbon. Why is it such a widespread element?

Recall that an element with less than 4 electrons in its outer energy level tends to lose them to another element which has more than 4. Conversely, an element with more than 4 electrons tends to take on electrons, rather than give them up. The most plentiful isotope of carbon has an atomic weight of 12, having 6 neutrons and 6 protons. In order to remain electrically neutral it must therefore have 6 electrons. With 2 in the first energy level, this leaves 4 for the second. Carbon thus stands in a quite unusual position. Instead of giving up or taking on 4 electrons in order to balance the outer energy level, carbon *shares* its electrons with other elements with which it forms chemical compounds.

When sodium chloride is formed by the uniting of sodium and chlorine, the sodium gives up an electron to the chlorine. Thus it takes on a positive electrical charge. The chlorine, by gaining an electron, becomes negatively charged. Such bonding is called *ionic.* Carbon, by virtue of its sharing, rather than giving up or taking on electrons, still remains essentially neutral. This sort of bonding is *covalent.* In general, due to their carbon atoms, organic compounds are covalent, while inorganic ones are ionic. It should be stressed, however, that there is no such thing as a 100 percent covalent or ionic compound. It is far better to speak of a compound as being *mostly* covalent or *mostly* ionic, whichever the case may be.

An example of the sharing ability of carbon can be seen in the formation of the organic compound *methane,* CH_4.

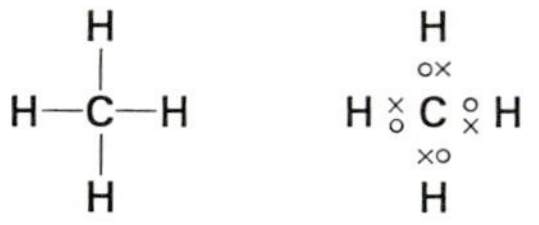

```
   H           H
   |           ox         Key:  o electron from carbon atom, C
H—C—H      H x C o H           × electron from hydrogen atom, H
   |         o   x
   H           xo
               H
```

The lines between the carbon and hydrogen atoms represent shared electrons. The diagram on the right is another way of representing this. The electrons are symbolized by small open dots and x's. Both methods of representation are widely used, but the single line technique will be used in this book from now on, unless it is necessary to show where each electron came from. *Each line represents a pair of shared electrons.* In the above example of methane, each hydrogen atom has one electron in its outer energy level, while carbon has four. The sharing of electrons enables both elements to complete their outer energy levels.

A unique feature of carbon is its formation of long chain molecules by sharing electrons with other carbon atoms. Methane is the first of a group of compounds making up a hydrocarbon *paraffin* series. The word "paraffin" is from the Latin and means "little affinity." The paraffin compounds are quite unreactive toward chemically active substances. Even highly reactive hydrofluoric acid, which etches glass, can be stored in bottles made of paraffin.*

---

* The compounds contained in the paraffin series do not normally appear in living organisms. However, their relative simplicity makes them good examples to demonstrate the versatility of the carbon atom in forming large and complex molecules.

If one hydrogen is removed from the methane molecule and a carbon atom takes its place, openings are created for more hydrogen atoms to become attached.

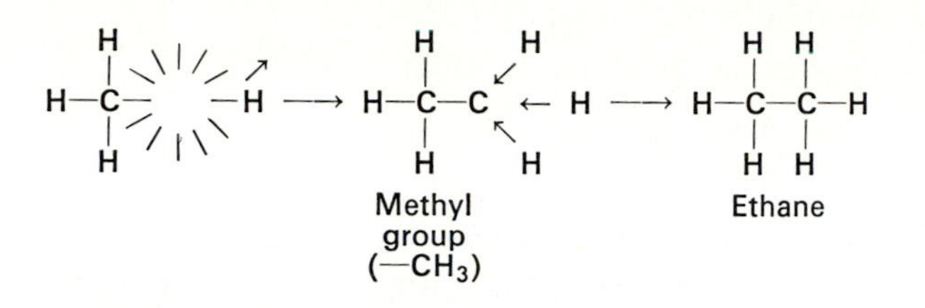

This new compound is *ethane,* C_2H_6, the second compound in the paraffin series.

Removal of one or more hydrogen from ethane, followed by the addition of another carbon atom plus 3 H atoms, results in a third compound, *propane,* C_3H_8.

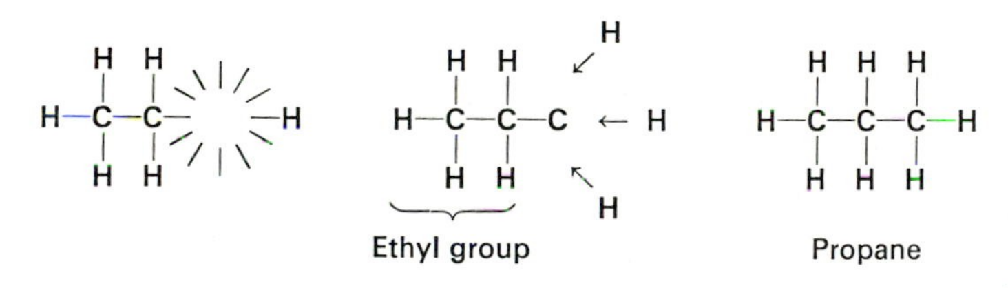

This seems as if it might be a never-ending process, and it almost is. The similar addition of other carbon atoms gives the following compounds:

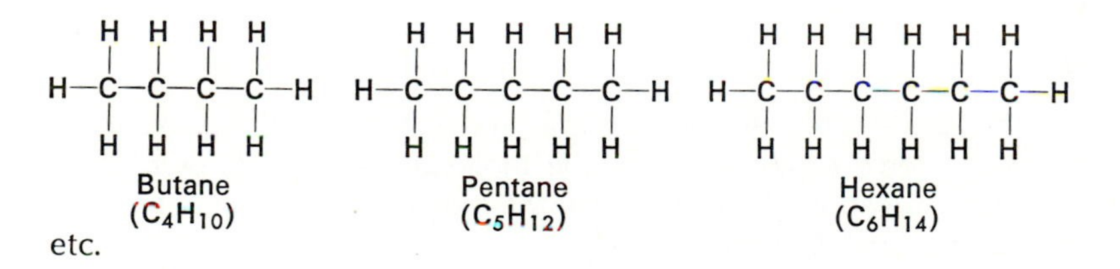

etc.

As can be seen, this ability of carbon to make chains can lead to the formation of a great many compounds. There is certainly no requirement that all the elements which unite with carbon must be hydrogen. For example, methane can be transformed into *methyl alcohol* as follows:

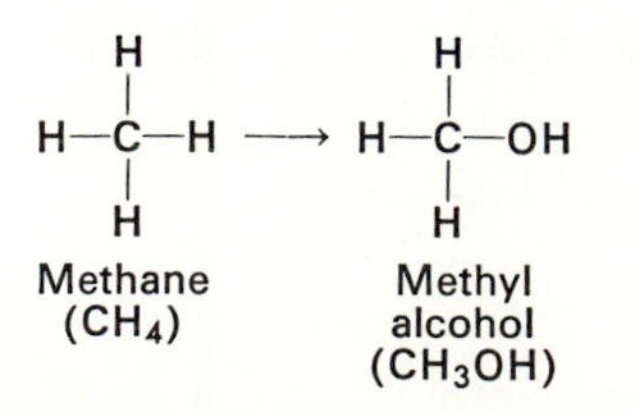

or to *trichloromethane,* better known as *chloroform.*

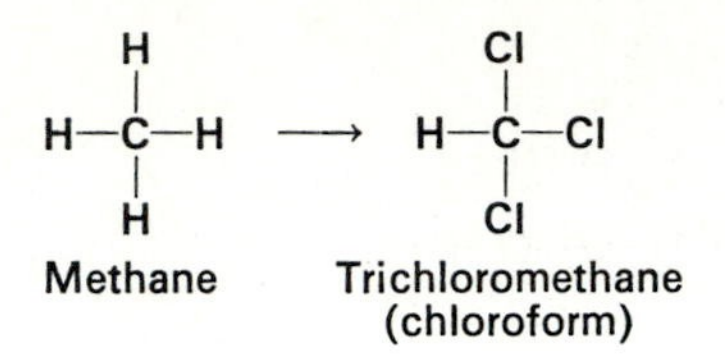

There is yet another factor which makes the theoretical number of carbon compounds infinite. The sharing of electrons between carbon atoms is by no means limited to straight chains. A chain of carbon atoms may turn corners, double back, and produce quite a variety of geometrical shapes. One of the simplest examples can be found in butane, C_4H_{10}, the fourth compound in the paraffin series. Normal butane (*n*-butane) is a straight-chain compound.

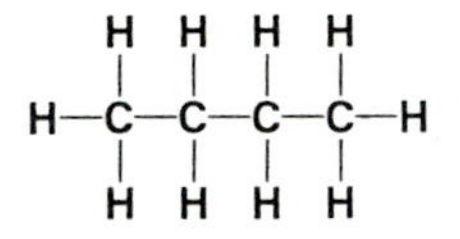

However, there is another butane known in which the atoms have the following arrangement:

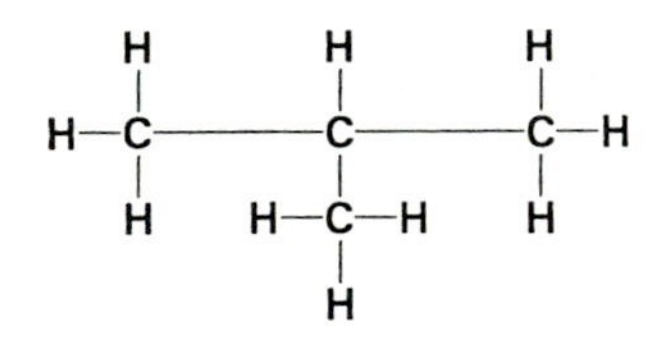

Note that this still has the same molecular formula as *n*-butane (C_4H_{10}). However, the shape of the molecule is quite different, and in organic compounds, *the shape of the molecule may well determine its physical and chemical properties.*

When two molecules have the same molecular formula, but different arrangements of atoms, they are called *isomers.* This second compound is an isomer of *n*-butane, and is called *isobutane.* It differs from *n*-butane in some of its physical and chemical characteristics. For example, *n*-butane releases more heat when burned than isobutane. Other differences are found in their boiling points, density, etc.

Another example of isomerism can be shown by starting with propane, C_3H_8.

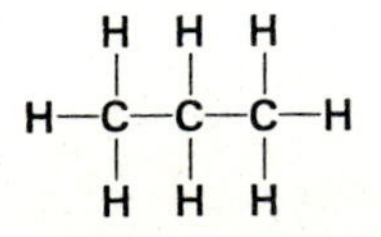

Removing the end hydrogen, and adding an —OH group, gives *n-propyl alcohol* (C_3H_7OH).

```
    H  H  H
    |  |  |
 H—C—C—C—OH
    |  |  |
    H  H  H
```

If, on the other hand, the hydrogen is removed from the middle of the molecule, and the —OH group put there, *isopropyl alcohol* (also C_3H_7OH), results.

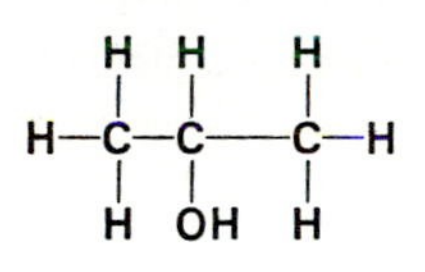

Sometimes, the carbon chains join ends to form a ring. *Cyclohexane* (C_6H_{12}) has such a ring-shaped molecule.

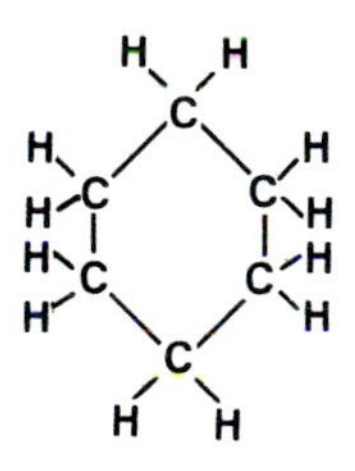

A better known compound with a ring-based molecule is *benzene,* C_6H_6, occasionally used as a cleaning fluid.

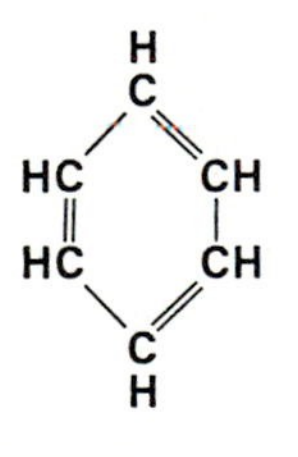

Here note that carbon forms double bonds with other carbon atoms. Recall that a single line represents the sharing of a single pair of electrons. A double line indicates that two pairs of electrons are being shared by the atoms involved. A comparison between a single and double bond, in terms of electron sharing, can be shown as follows:

C ° × C	C ×° °× C	Key: ° Electron from one carbon atom
Single bond	Double bond	× Electron from other carbon atom

When double bonding occurs, there is room for more atoms to be added to the compound at the double-bond location. To illustrate this, look at the structural formula of ethane below.

```
   H  H
   |  |
H—C—C—H
   |  |
   H  H
```

Here, all of the bonds are filled. Such a compound is said to be *saturated*.* However, if 2 hydrogen atoms are taken away, this leaves spaces to be filled. If no other elements are available, a double bond forms between the carbon atoms from which the hydrogen atoms were removed. This indicates two pairs of electrons are being shared at this point. This forms the compound *ethene*, C_2H_4.

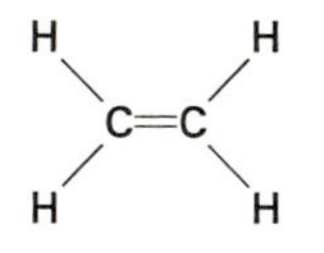

The removal of 2 more hydrogen atoms leads to a triple bond between the carbon atoms, forming the compound *acetylene*, C_2H_2, well known for its use in blowtorches. Here, three pairs of electrons are being shared.

$HC\equiv CH$

As might be expected, a saturated compound is chemically inert and unreactive as far as the addition of more atoms is concerned. There are no more electrons available to be shared with other elements. Unsaturated compounds, on the other hand, are generally quite reactive as far as addition of other atoms is concerned, and tend to go toward the saturated state.

In ethene (commonly known as ethylene), the presence of the double bond makes the compound very reactive. Double-bonded ethylene can become completely unreactive, simply by letting go of one pair of shared electrons. This makes an electron available at either end for the addition of other electron-sharing carbon atoms. The chain molecule, *polyethylene*, is thus formed. Polyethylene's lack of chemical reactivity is shown by the fact that it is widely used for tubing and containers which must convey or hold highly reactive chemicals (Fig. 7–1).

From the preceding, you can see the tremendous possibilities that exist for the formation of different carbon compounds. The fact that

---

* A saturated carbon chain can take on no more elements, since all of the available electrons are being shared.

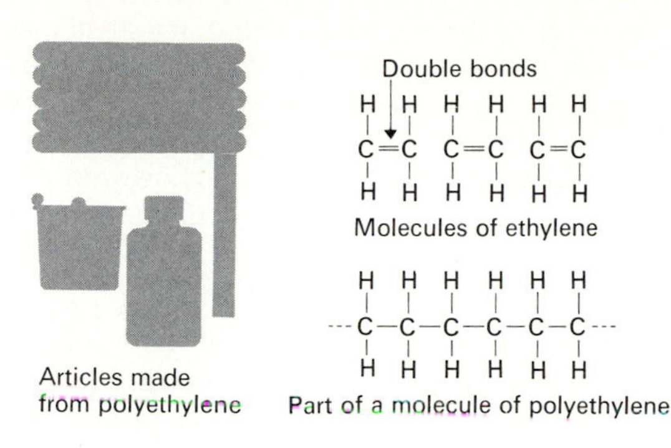

7–1
Polyethylene (polythene), produced from ethylene, is a very unreactive substance. It therefore has uses in situations where this quality is desirable.

carbon can share its electrons with other carbon atoms, and thus produce long carbon chains, accounts for many of the interesting characteristics which we associate with living material.

However, carbon is not the only element found in living organisms which can share electrons. Let us re-examine the factors which influence the formation of a compound from various atoms.

Atoms will unite to form a compound if, in doing so, the outer electron shells assume a stable arrangement. A stable arrangement, recall, is one in which the outer electron shells contain the proper number of electrons. For example, the proper number for carbon is 8. For hydrogen, it is 2. In methane, the sharing of the electrons from 4 hydrogen atoms completes the outer shell of carbon, since it already has 4 of its own. At the same time, each of the outer shells of the 4 hydrogen atoms become complete, with one electron of each hydrogen atom added to the one contributed by carbon.

With this in mind, we can see that any atom with 4 or more electrons in its outer shell can complete its octet by sharing these electrons with other atoms. Such sharing will result in a covalently bonded chemical compound if all of the atoms involved assume a stable electron configuration, i.e., a complete outer electron shell.

Let us take a specific example. Oxygen, with an atomic number of 8, has 2 electrons in its *K* shell and 6 in its *M* or outer shell. These electrons are shown below as two electron pairs and two single electrons.

```
 ××
× O ×    (Oxygen)
 ××
```

For oxygen to form a compound, two more electrons must be added to the outer shell to form a stable octet arrangement. Our first thought might be to transfer two electrons to oxygen from some other atom, thereby filling the outer shell. Such a situation exists in compounds such as sodium oxide, Na_2O. Here, each of two sodium atoms donates its lone outer-shell electron to oxygen. The result is a stable electron configuration for both atoms.

Still another possibility exists, however. The outer shell of oxygen could be completed by one or more atoms which could share two electrons with oxygen. A good example of this is the sharing of electrons between oxygen and hydrogen.

H ×° O °× H or H—O—H (Water)

Such an arrangement completes the outer shell of both oxygen and the two hydrogen atoms. The atoms which contribute the electrons for sharing with oxygen do not have to be hydrogen. Oxygen, for example, can share just one electron with hydrogen. This results in an outer shell with seven electrons, one short. The eighth electron can be contributed by another element, such as carbon.

Another problem remains, however. Although the outer shell of oxygen is now complete, the carbon still needs three electrons. These can be obtained by electron sharing with three hydrogen atoms.

H C O H (with H above and below C) or H—C—O—H (with H above and below C)

Here, by electron sharing, a compound is formed from carbon, hydrogen, and oxygen.

Two oxygen atoms are joined by two covalent bonds to form an oxygen molecule:

O ×× ×× O or O=O

The double covalent bond satisfies the requirement for eight electrons in the outer electron shell. The unpaired electrons give rise to the chemical properties of elementary oxygen.

Carbon, of course, can share two electrons with an oxygen atom:

Such a situation produces a double bond between the carbon and oxygen atom.

These examples become of major importance when considering compounds in living material. Many contain only carbon, hydrogen, and oxygen. Most are formed by the electron sharing just illustrated.

Other atoms can be examined for their possibilities of forming compounds by electron sharing. A good one is nitrogen. This atom has five electrons in its outer shell.

```
 □
□N□
□□
```

A moment's reflection on electron-sharing "rules" should make such compounds as the following quite logical.

```
      H                      H
     □○                      |
   H○□N□○H       or      H—N—H          (Ammonia)
     □□

  H    H                 H   H
  ×○   □○                |   |
H○×C×□ N□×H      or    H—C—N—H          (Methyl amine)
  ○×   □□                |
  H                      H
```

Compounds containing nitrogen are of considerable biological importance. They will be considered later in the discussion of amino acids and proteins.

7–3 THE DETERMINATION OF STRUCTURAL FORMULAS

How can the organic chemist tell the shape of a molecule which he cannot see? Furthermore, since both *n*-butane and isobutane are C_4H_{10}, how can he tell which shape goes with which structural formula?

It is beyond the scope or needs of this book to describe in detail all of the techniques used to determine structural formulas. On the other hand, a very brief look at these techniques may serve to give more of a feeling of familiarity and understanding of the formulas which are necessarily encountered in an introductory study of biology.

Several kinds of information are used to determine structural formulas. First, we must find out by chemical analysis what elements are present in the substance and in what proportions. Second, we can determine certain chemical and physical properties (such as boiling and freezing points) of the substance. These properties can then be compared to those of other known compounds. Third, samples of the unknown are allowed to react with other known chemicals, and the products of these reactions studied. Some of the products may be fragments of the unknown molecule which give clues as to its chemical make-up. Suppose, for example, the unknown compound gives evidence of containing a $-CH_3$ or methyl group. The unknown can then be tested with a chemical compound known to give methane in the presence of a methyl group. If methane appears, the hypothesis concerning the presence of a methyl group in the unknown compound is supported.

When the organic chemist, using such techniques, feels that he has a fairly correct idea of the structure of the unknown compound, he is then ready to test his hypothesis. Synthesis of the unknown molecule is attempted, using reactions which lead to a product with the hypothesized structure. If the final product of this synthesis has physical and chemical properties which are identical with those of the unknown, the hypothesis concerning its structure is firmly supported.

In the past few years, the development of highly refined instruments has enabled the organic chemist to determine with great accuracy the precise spatial positioning of atoms and groups of atoms within molecules. One of the most helpful of these techniques is *x-ray diffraction*. While x-ray diffraction is a complex process and calls for a great deal of mathematical analysis, its principle is fairly simple. A beam from an x-ray tube hits the unknown sample and is then deflected onto a photographic plate. A certain pattern of dots is recorded on this plate. The distances between the dots are a measure of the interatomic dimensions of the unknown molecule. Using this information, the types of atoms and their location within the molecule can be determined. This is *somewhat* similar to determining the shape of an object from the shadow it projects when exposed to light, but the process is a great deal more complicated than this.

Another technique is the use of *ultraviolet* or *infrared spectrophotometry*. Literally translated, this means "measurement with the light spectrum." The light spectrum, of course, is made up of the various bands of light. These, in turn, differ as to wavelength. *Different wavelengths of light are absorbed by different atoms or groups of atoms.* This fact is the basis of spectrophotometry.

In infrared spectroscopy, for example, molecular structure is determined by studying the molecule's selective absorption of wavelengths falling within the infrared portion of the spectrum. One wavelength may be completely absorbed, a second only partly, while a third may not be absorbed at all. Since specific absorption bands have been linked to the presence of certain structural units, identification of unknown molecular groups is thus possible.

While the physics involved in spectrophotometry is complex, there is nothing particularly difficult about understanding its principle. It simply involves the association of one physical phenomenon with the presence of another. In this case, it is the observance of the presence of a certain spectral appearance on an instrument with the presence of a certain group and arrangement of atoms. In a similar way, the recognition that certain tastes and odors go with certain substances is learned by association.

The determination of structural formulas for biologically important molecules will become of even greater importance in the future. To date, the three-dimensional structures of only a few biologically important macromolecules have been worked out. In 1962, M. H. F. Wilkins, J. D. Watson, and F. H. C. Crick won the Nobel Prize in Medicine for their determination of the structure of deoxyribonucleic acid (DNA). In the same year, J. C. Kendrew and M. F. Perutz shared the Nobel Prize in Chemistry for their determination of the molecular structure of the muscle protein, *myoglobin*. The blood protein hemoglobin, very similar to myoglobin, has also been studied and its three-dimensional structure determined for a few species. In addition the structure of the enzymes lysozyme and ribonuclease have been completely determined to date; and the structure of dogfish lactic dehydrogenase (also an enzyme) is virtually completed. X-ray diffraction played a major part in all these investigations.

7–4 THE FORMULAS OF ORGANIC CHEMISTRY

A biology student encounters many organic compounds. Some of these have very complex formulas. It is necessary, therefore, that the student be familiar with the written language of the organic chemist. With such a background, he can more fully understand the nature of the biologically important compounds with which he will deal, from simple carbohydrates to highly complex nucleic acids and nucleoproteins.

Often, the organic chemist simply uses *molecular* formulas, with which most people are familiar. $C_6H_{12}O_6$ is an example of a molecular formula. Such molecular formulas indicate the kinds of atoms in the compounds, the precise number of each, and can be used to determine molecular weight. However, just as important to a compound as the number and the kinds of atoms in the molecule is the way in which they are arranged. This is particularly true with organic compounds. Therefore, the organic chemist uses *structural* formulas, which indicate the *geometric shape* of the molecule. These structural formulas may be as elaborate or as simple as the occasion demands. A highly illustrative picture of the arrangement of the atoms in the ring-shaped molecule of glucose may be given as follows:

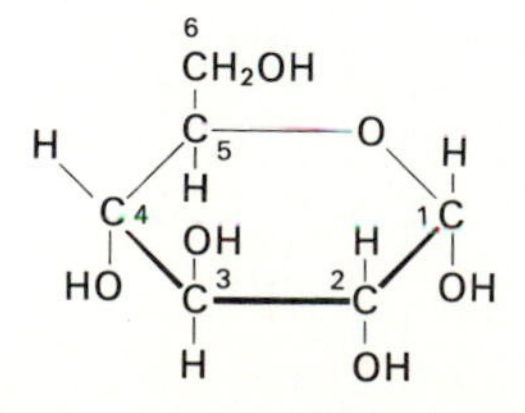

Each carbon atom is numbered for easy identification of places at which various other atoms or groups of atoms may attach. The carbon atoms are imagined to lie in a plane that protrudes at a right angle from the page. The H and OH groups then lie above or below this plane. The geometric structure of glucose shown here is only one of several which can exist in equilibrium with each other in the aqueous medium of the cell. For the sake of convenience, however, we will refer to glucose by the planar configuration described above.

Less time-consuming to create and more commonly seen is the same formula given in 2 dimensions. In this case, the same form of glucose which is given above is written as:

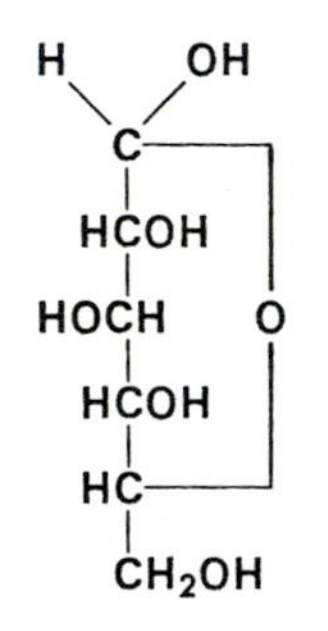

This structural formula is slightly less informative than the first one. The relative positions of the atoms are indicated, but it is left to the observer to make the transition from 2 to 3 dimensions.

Often, in writing complex organic structural formulas, certain shortcuts are employed. Rather than writing down all of the atoms, the organic chemist has adopted a shorthand method, using only bond lines. He eliminates writing atoms which occur repetitively and are regular components of the basic molecular framework. For example, the complete benzene molecule is as follows:

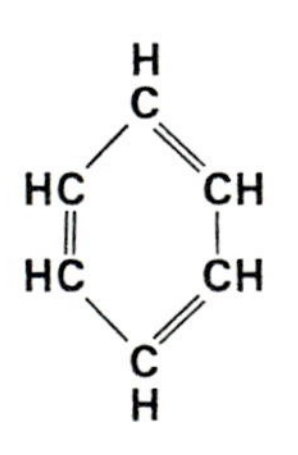

However, the organic chemist symbolizes benzene by simply drawing its ring:

Thus, the formation of such a compound as *methyl benzene* (*toluene*) by removal of a hydrogen atom and attachment of a methyl ($—CH_3$) group, could be written as:

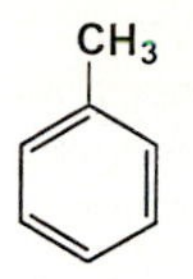

Not only is this easier to write, it also focuses attention on the significant part of the molecule. This system of symbolization may become quite complicated in itself, but bear in mind how tedious some formulas would be to write without it. The rather simple structure of *ascorbic acid* (vitamin C), as pictured in a ball-and-stick model, may be compared with a full structural formula in the middle and a shorthand formula to the right.

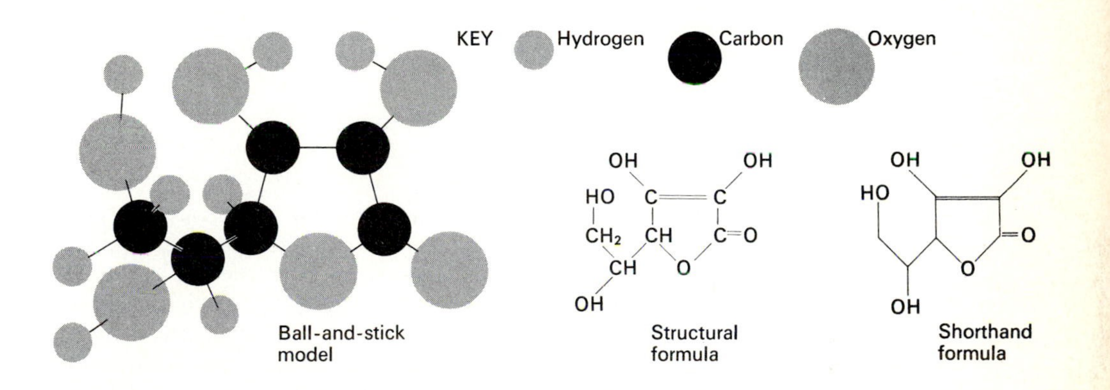

Some compounds have an arrangement which permits an even simpler manner of representing their structural formulas. For example, hexane, C_6H_{14}, may be written structurally as:

```
    H  H  H  H  H  H
    |  |  |  |  |  |
 H—C—C—C—C—C—C—H
    |  |  |  |  |  |
    H  H  H  H  H  H
```

But it may also be written as:

$CH_3CH_2CH_2CH_2CH_2CH_3$

or simply:

$CH_3(CH_2)_4CH_3$

A moment's study shows that these latter two representations tell the same story as the first formula above and are often more convenient. They cannot always be used, however. Often, the structural formula must be used, despite its taking more space, in order to convey the full amount of information known about the molecule.

There are still other more elaborate means of writing structural formulas to which only passing attention will be given. For example, 2 hydrogen atoms may be removed from the center carbon atoms of the hexane molecule and replaced with a $—CH_3$, or methyl group, as follows:

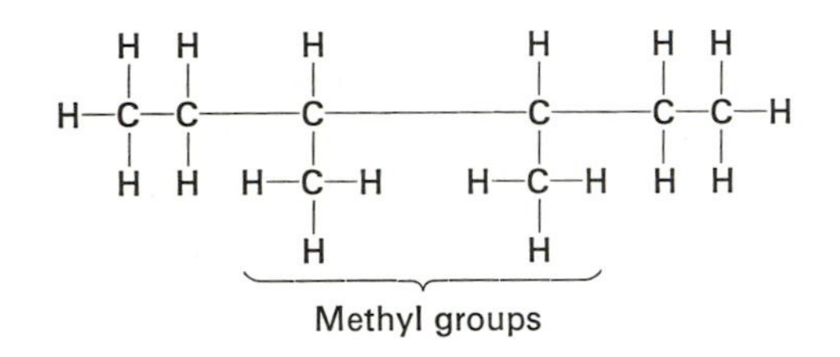

This may also be written as:

$$\begin{array}{l} CH_3CH_2CH—CH—CH_2CH_3 \\ \qquad\quad\;\; | \qquad\; | \\ \qquad\quad CH_3 \;\; CH_3 \end{array}$$

Finally, it may be noted that this last compound is known as *3,4 dimethylhexane*. Organic chemistry has a rather elaborate nomenclature, which is highly descriptive. In this particular name, the *hexane* refers to the basic 6-carbon hydrocarbon to which the side groups are attached. The *dimethyl* means that there are 2 methyl groups on the molecule. The 3,4 refers to the numbers of the carbon atoms to which the methyl groups are attached.

That this descriptive nomenclature is handy should be quite obvious. An organic chemist can hear the name of a compound, say *4-isopropyl-6-n-propyl-3, 7-diethyl nonane,* and immediately reconstruct its structural formula as follows:

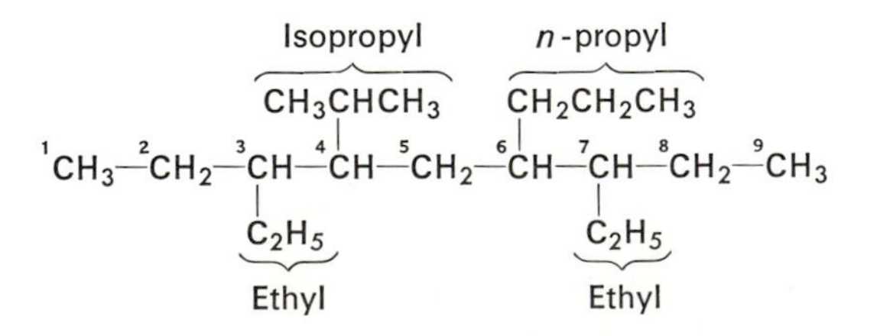

The small numbers going from one to nine indicate the particular carbon atom to which various side groups are attached. Thus, the isopropyl group is attached to the 4th carbon; the n-propyl group to the 6th carbon; the two ethyl groups are attached to the 3rd and 7th carbons. The final word, nonane, indicates that there are 9 carbon atoms in the longest chain of the molecule.

As can be seen, the system can become quite complicated, though once mastered it is really rather simple. There is certainly no need for the reader to learn it now. However, this exposure to the methodology of formula-writing may make less appalling the names of such biologically important compounds as *dichlorodiphenyltrichlorethane*—more commonly known as DDT.

7–5 SOME FUNDAMENTAL ORGANIC GROUPS

Although there are an almost endless number and variety of carbon compounds, there are a few building units which occur repeatedly. These units not only confer similar properties on otherwise quite dissimilar compounds, but also provide a means of identifying and describing these compounds. We call these groups *organic radicals,* or *functional groups*. Table 7–3 lists some of the functional groups most frequently encountered in biochemical work. These groups play important roles in organic compounds. The alcohol group is found in alcohol compounds, various sugars, and other carbohydrates. All of these compounds are intermediary or end products of cellular metabolism. (See Fig. 7–2.) In many cases the addition or removal of an alcohol group completely changes the chemical nature of an organic compound.

There are two types of carbonyl groups. Aldehyde groups contain a central carbon atom double-bonded to oxygen and single-bonded to hydrogen. Aldehydes are found as intermediate products in such biological processes as photosynthesis and respiration. Like alcohol groups, the aldehyde groups are found chiefly in carbohydrates and their products. The compound *formaldehyde,* used as a preservative for biological specimens, has the formula

```
   H
   |
H—C=O
```

Formaldehyde prevents the breakdown of protein by inhibiting the growth of bacteria.

The second type of carbonyl groups are *ketones*. Ketone groups have their central carbon atom double-bonded to oxygen and joined to two other carbons by a single bond to each. Ketone groups occur in intermediate compounds, in the breakdown or synthesis of many amino acids (the building blocks of proteins), as well as in the metabolism of fats.

The carboxyl group (COOH) is one of the most important functional groups. It is found in amino acids, fatty acids, and many other compounds. The carboxyl group is also called the acid group because in solution it tends to ionize and release protons (see Eq. (6–2), Section 6–2). Carboxyl groups are what give amino and fatty acids their "acid" characteristics.

(Continued on page 134.)

THE SHAPE OF MOLECULES

Table 7–2 was prepared to emphasize the effect of shape on certain molecules that were mentioned in the preceding pages. These molecules are generally *not* biological molecules. Examine the differences in shape and physical and chemical properties.

Note in the table that the molecules with the same molecular weight but different structures have different melting and boiling points. This demonstrates that the internal arrangement of atoms in a molecule will influence the attractive forces between it and identical molecules around it in a particular state of matter (gas, liquid, or solid).

TABLE 7–2
The shape of molecules

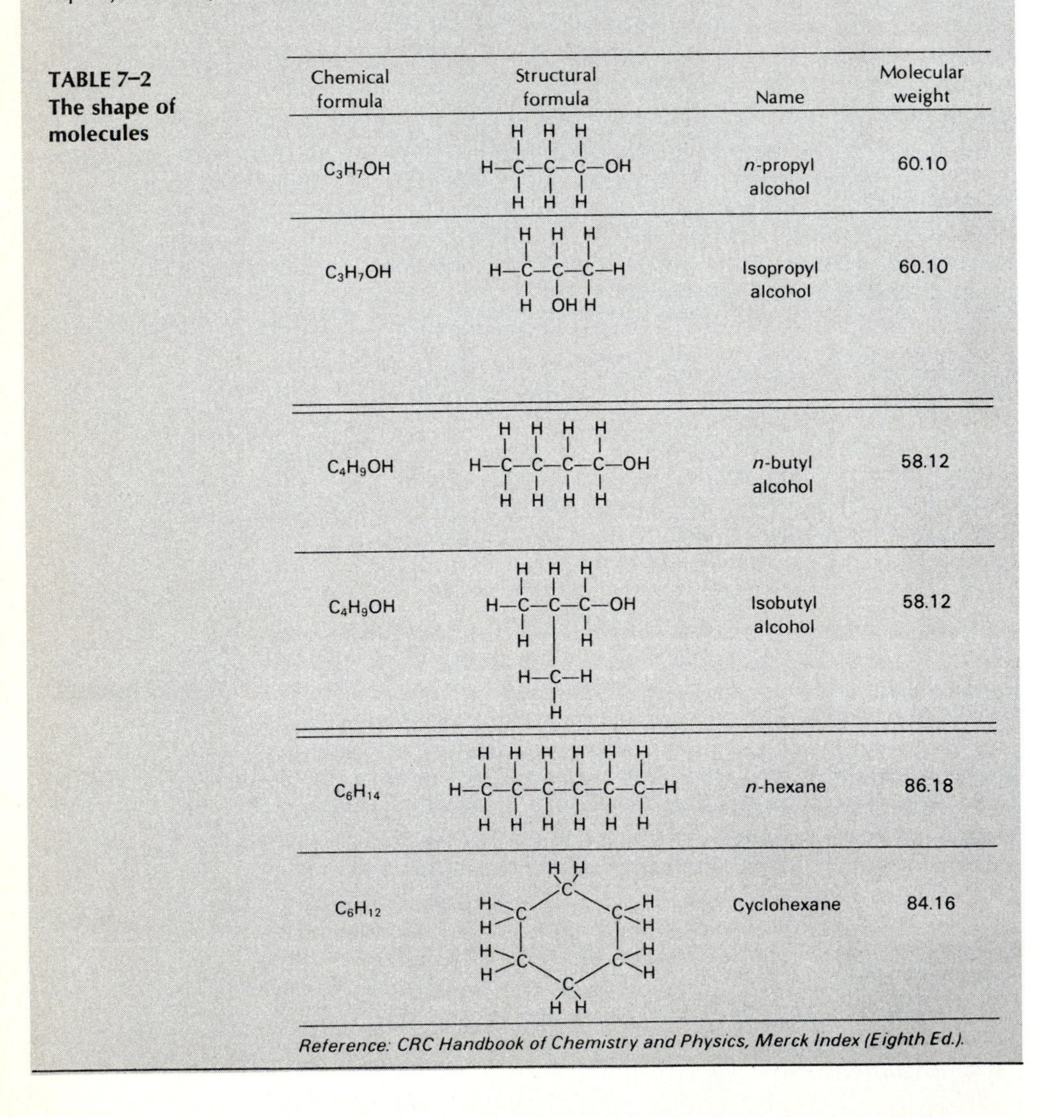

Chemical formula	Structural formula	Name	Molecular weight
C_3H_7OH	H H H H—C—C—C—OH H H H	*n*-propyl alcohol	60.10
C_3H_7OH	H H H H—C—C—C—H H OH H	Isopropyl alcohol	60.10
C_4H_9OH	H H H H H—C—C—C—C—OH H H H H	*n*-butyl alcohol	58.12
C_4H_9OH	H H H H—C—C—C—OH H H H—C—H H	Isobutyl alcohol	58.12
C_6H_{14}	H H H H H H H—C—C—C—C—C—C—H H H H H H H	*n*-hexane	86.18
C_6H_{12}	H H C H H—C C—H H H H—C C—H H C H H	Cyclohexane	84.16

Reference: CRC Handbook of Chemistry and Physics, Merck Index (Eighth Ed.).

A solid has its atoms arranged in a fixed, orderly fashion, forming a three-dimensional pattern composed of repeating pattern units throughout the entire structure. In a liquid, the atomic or molecular arrangement varies rapidly with time. In gases, there is complete disorder in the molecular arrangement.

In going from a solid to a liquid state, energy must be supplied to permit the greater freedom of movement in the liquid compared to the solid. The temperature at which the solid and liquid phases are in equilibrium is called the *melting point.*

Boiling involves the escape of molecules from the liquid. Enough energy must be supplied to overcome the attractive forces between the molecules in the liquid so that they may escape as gas. The temperature at which the liquid and gaseous phases are in equilibrium is called the *boiling point.*

Melting point (°C)	Boiling point (°C)	Some uses
−127.0	97.2	Used as solvent for resins and cellulose esters
−89.5	82.4	Used as rubbing alcohol; in antifreeze compounds; solvent for gum, shellac, essential oils; in quick-drying inks; in quick-drying oils; in hand lotions, aftershave lotions, and similar cosmetics
−90	117–118	Solvent for fats, waxes, resins, shellac, varnish, gums; manufacture of lacquers, rayon, detergents; in microscopy for preparation of paraffin imbedding material
−108	108	Manufacture of esters for fruit-flavored essences; solvent in paint and varnish removers
−95.3	68.7	Used to determine refractive index of minerals; filling for thermometers instead of mercury, usually with blue or red dye
6.5	80.7	Solvent for lacquers and resins; paint and varnish removers; in extraction of essential oils; in manufacture of solid fuel for campstoves

The amino group is found primarily in amino acids, though it occurs in other organic compounds as well. In solutions with a high concentration of protons, amino acids have a tendency to take on additional protons and form an ammonium (RNH_3^+) ion. The amino group thus acts as a base.

The sulfhydryl group consists of a sulfur atom bonded to a hydrogen atom. It is widely found in proteins, particularly enzymes. Certain compounds which block enzyme activity have been experimentally shown to perform this task by combining with the sulfhydryl groups of the enzyme. This strongly suggests a major role for the sulfhydryl group in the chemistry of enzyme action.

TABLE 7–3 Some organic functional groups

Name of functional group	Structural formula	Structural formula of typical molecule in which functional group may be found*	
1. Hydroxyl or alcohol	R—C(R)(R)—OH	H—C(H)(H)—[C(H)(H)—OH]	Ethyl alcohol
2. Carbonyl			
(a) Aldehyde group	R—C(H)═O	H—C(H)(H)—[C(H)═O]	Acetaldehyde
(b) Ketone group	R(R)C═O	H—C(H)(H)—[C(═O)]—C(H)(H)—H	Acetone
3. Carboxyl	R—C(═O)—OH	H—C(H)(H)—[C(═O)—OH]	Acetic acid
4. Amino	R—N(H)—H	[H—N(—H)]—C(H)(H—C(H)—H)—C(═O)—OH	The amino acid alanine
5. Sulfhydryl	R—S—H	H—N(—H)—C(H)(H—C([SH])—H)—C(═O)—OH	The amino acid cysteine

* The functional groups are outlined in dashed lines. The R stands for hydrogen or whatever group of atoms is attached to the functional group.

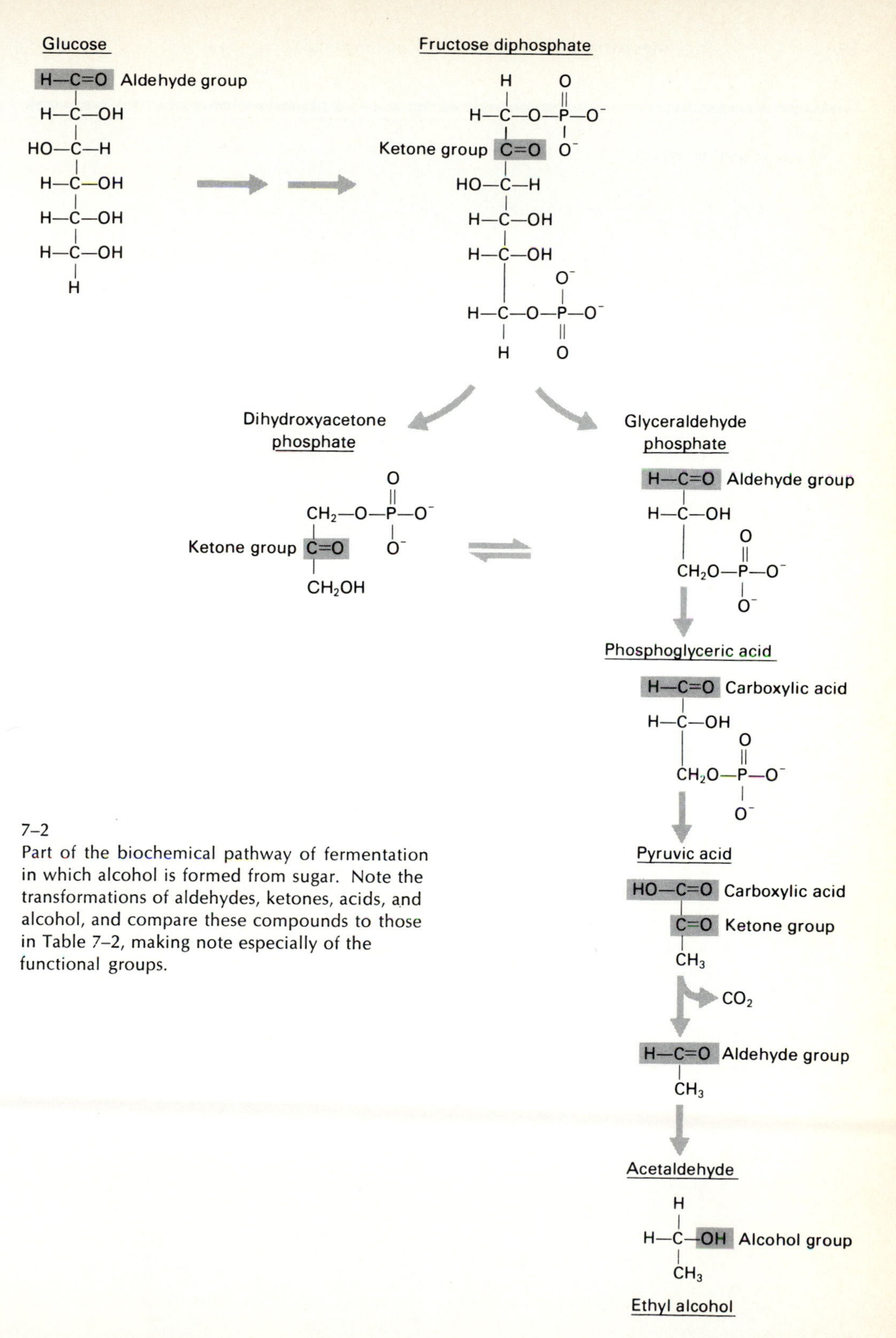

7–2
Part of the biochemical pathway of fermentation in which alcohol is formed from sugar. Note the transformations of aldehydes, ketones, acids, and alcohol, and compare these compounds to those in Table 7–2, making note especially of the functional groups.

SULFHYDRYL POISONS

A number of enzymes contain sulfhydryl (SH) groups that are necessary for catalytic activity. These enzymes can be "blocked" or "poisoned" by substances that react chemically with the SH groups. Two well-known examples of sulfhydryl poisons are the heavy metals lead and mercury, which inhibit enzymes in proportion to their concentration. If the enzymes are inhibited more rapidly than they are synthesized, the "poisonous" effects will be observed.

Lead is often a component of outside paints, storage batteries, gasoline additives, unglazed china, ceramics, insecticides, plumbing, lead toys, solder, foil wrapper, and bullets. Lead poisoning is usually slow due to chronic (long duration, frequent recurrence) exposure, and its effects accumulate until a "poisonous" level is reached. Children are especially susceptible because their growing tissues tend to accumulate the material more rapidly and because, proportionately, a given dose has a greater effect on a child than on an adult. The symptoms of lead poisoning, that is, the gross effects of reactivity with sulfhydryl groups, are metallic taste, nausea and vomiting, abdominal pain, diarrhea, anorexia (loss of appetite), and, in severe cases, liver or kidney damage, central nervous system damage, encephalopathy (convulsions), partial paralysis, coma, and death.

Mercury is found in certain disinfectants, antiseptics, insecticides, seed and grain preservatives, paints, laboratory instruments, and dry-cell batteries, and is used for other purposes. It can sometimes vaporize enough at room temperature to reach toxic levels. As with lead, mercury poisoning is also cumulative (accumulates with each additional exposure). In chronic low-level exposure, the kidneys will eliminate mercury for a long time (months) after the last exposure without any apparent damage, but in acute (severe, high-dose) exposure, symptoms occur within minutes —a burning sensation in the mouth and throat, nausea, vomiting, severe abdominal pain, diarrhea, extreme thirst and salivation, rapid weak pulse, cold moist skin, slow breathing, shock; perhaps death. Chronic symptoms include inflammation of the mouth, swollen salivary glands, soft spongy gums, blue-black gum line, excessive salivation, metallic taste, foul breath, muscular tremors, and mental and nervous symptoms. One of the most famous cases of mercury poisoning involved the town of Minamata, Japan, whose waters were polluted by a chemical plant and mercury entered humans through the fish.

Source: Kaye, Sidney, *Handbook of Emergency Toxicology* (Third Ed.). Charles C. Thomas, Springfield, Ill., 1970; and White, Abraham, Philip Handler, and Emil L. Smith, *Principles of Biochemistry* (Fifth ed.), McGraw-Hill, New York, 1973, pp. 239–240.

Chapter 8
Some Fundamental Organic Substances in Living Material

PART 1
CARBOHYDRATES

8-1
INTRODUCTION

Living organisms are composed of considerable numbers of different atoms constantly being exchanged, one for another. These atoms are united to form many types of molecules, all playing important roles in the functioning of the organism. It is difficult to select any one group of these molecules and say that they are the "most important." It might possibly be justifiable to call the engine of an automobile more important than the ignition key, yet both are essential for proper functioning.

So it is with the chemistry of living things. In the clotting of blood, for example, fibrinogen plays a vital role in the formation of the clot protein, *fibrin*. Yet were it not for the presence of many other inorganic and organic buffers, the pH would be such that the fibrinogen would not have a chance to play its leading role.

This section presents some physical and chemical properties of compounds which are among the most fundamental in living organisms. A distinction should be made between the words "fundamental" and "most important." In our analogy of the car, it will probably be agreed that the engine is certainly more fundamental than the ignition key, since the car could hardly keep going without it. Therefore, it merits more attention in a discussion concerning the functioning of the automobile. This is precisely the justification made here for the choice of molecules now to be discussed.

8–2 TYPES OF CARBOHYDRATES

The carbohydrates exist in many different compound forms, the most important of which are given in Table 8–1. Several important features about carbohydrates can be seen from this table. First, observe that the hydrogen-to-oxygen ratio in these compounds is generally 2:1. This, of course, is the same ratio found in water. The *carbohydr-* portion of the word carbohydrate indicates the presence of carbon and hydrogen. The entire word, carbohydrate, implies a water-containing, or hydrated, carbon. Although, chemically speaking, this implication is incorrect, it does emphasize the numerical ratio of these

TABLE 8–1 Common carbohydrate compounds

$C_3H_6O_3$—triose sugar—example: glyceraldehyde.

$C_4H_8O_4$—tetrose sugar—example: erythrose.

$C_5H_{10}O_5$—pentose sugars—examples: ribose, deoxyribose,* ribulose, xylose, arabinose.

$C_6H_{12}O_6$—hexose sugars—examples: glucose, fructose, galactose, mannose.

$C_7H_{14}O_7$—heptose sugar—example: sedoheptulose.

The compounds above are all classified as simple sugars, or *monosaccharides.* The Latin *saccharum* means sugar and *mono*-means one.

$C_{12}H_{22}O_{11}$—examples: sucrose, lactose, maltose.

These are double sugars, or *disaccharides.*

$C_{18}H_{32}O_{16}$—example: raffinose.

This is a triple sugar, or a *trisaccharide.*

$(C_6H_{10}O_5)_n$—examples: starch, glycogen, dextrins, gum, mucilage, inulin, cellulose.

Since these compounds are composed of *n* number of monosaccharide units (the exact number varying from one to another) they are called *polysaccharides.*

* Deoxyribose, as its name implies, has less oxygen than ribose.

three atoms. By weight, oxygen, carbon, and hydrogen are the first, second, and third most plentiful elements in living matter.

Since the ratio of carbon, hydrogen, and oxygen atoms is so constant, a general or *empirical* formula can be devised that applies to almost all carbohydrates. This relationship is expressed by $(CH_2O)_n$.

Refer again to Table 8–1. Note that many of the examples have the same molecular formula. Glucose and fructose, for example, are both $C_6H_{12}O_6$. Since they contain the same atoms in the same proportions, why is any distinction made between them?

It has already been pointed out that in the chemistry of organic compounds, the shape or configuration of the molecule may greatly affect its physical and chemical properties. Therefore the arrangement of the atoms and groups of atoms must be studied as well as their proportions. Two piles of a child's blocks may have identical letters on them, but the way those letters are arranged is all-important to the message they convey. *Structural formulas* must therefore be used to show the spatial arrangement of the parts.

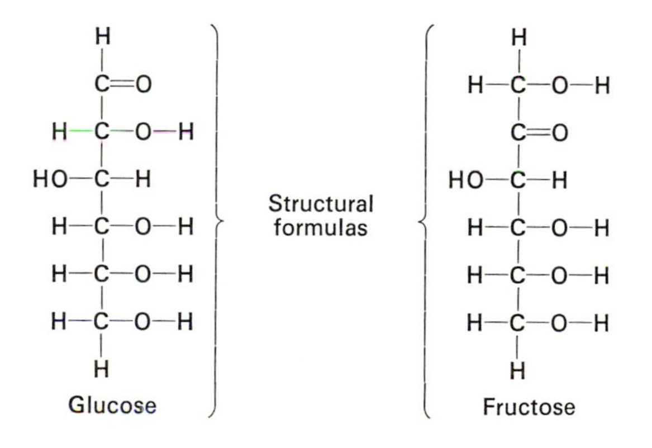

The differences between the molecules may seem to be slight, but they are nonetheless very important, and affect their chemical and physical characteristics. In taste, for example, glucose is not nearly so sweet as fructose.

As seen in Section 7–2, in cases where two or more compounds have the same molecular formula, but differ structurally, the term *isomer* is applied. Table 8–1 shows that sucrose, maltose, and lactose are isomers with the molecular formula $C_{12}H_{22}O_{11}$. Ribose and ribulose are isomers with the molecular formula $C_5H_{10}O_5$.

Note that the structural formula given for glucose in Section 7–4 differs from the one shown above. Glucose can exist in two molecular forms, a ring structure and a straight-chain form. Since, for living organisms, glucose is probably the most fundamental sugar, it is perhaps worthwhile to enlarge upon this point.

The ring structure of glucose is formed by an internal rearrangement of bonds that results in the linking of carbon #1 to carbon #5 through an oxygen atom. This is accomplished by a reaction with water as shown below.

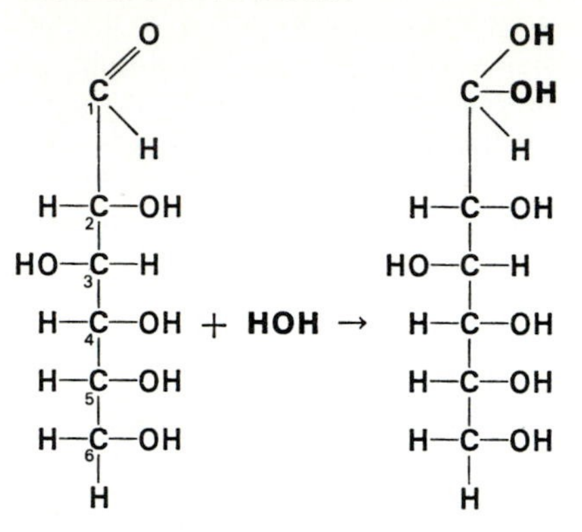

This is followed by the final stage of the reaction, which yields:

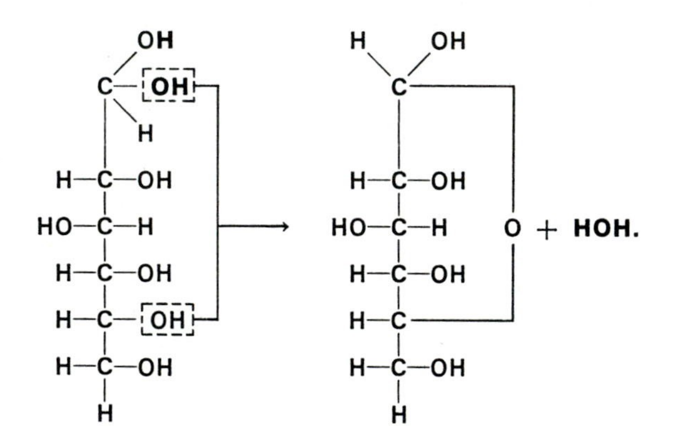

Actually, there are two possible forms which this ring-shaped molecule can take (Fig. 8–1). An equilibrium exists between these two forms so that in any glucose solution approximately 37 percent of the molecules have the alpha form and 63 percent the beta form.

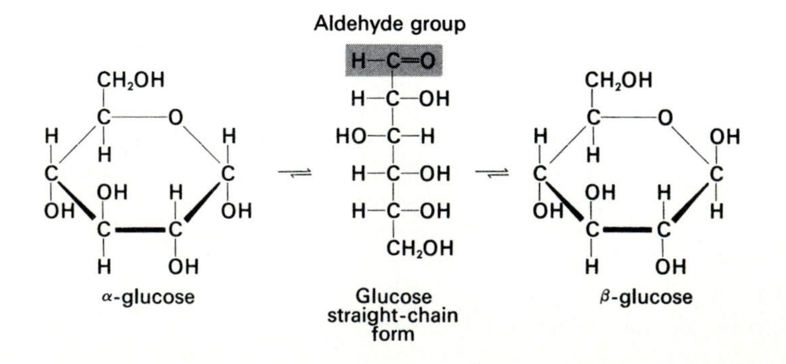

8–1
An equilibrium exists between the alpha and the beta forms of the ring-structured glucose molecule in solution. Note that the molecules pass through the straight-chain form to get from one ring structure to another.

Note that in solution the straight-chain form of glucose exists only as a transition compound in the change from the alpha to the beta forms of ring-structured glucose molecules. Only this straight-chain form leaves a carbonyl group (in this case, an aldehyde) exposed. *In the ring structure, this aldehyde group is eliminated* by the joining of the number one carbon atom to the number five carbon atom. Many tests for the presence of glucose rely upon the exposure of the aldehyde group. Fehling's solution, which contains copper sulfate, gives such a test. When Fehling's solution is placed in a warm solution of glucose, the sugar is oxidized and the Fehling's solution reduced. A burnt orange or reddish precipitate (copper oxide) is formed.*

If all of the glucose molecules in the solution being tested were in the straight-chain form, and thus had an aldehyde group exposed, the test reaction would be expected to proceed fairly rapidly. However, this is not the case. The slowness of the reaction indicates that only a few of the glucose molecules are in the open-chain state with an exposed aldehyde group. This, in fact, is just what we have demonstrated above. The Fehling's solution reacts only with the straight-chain molecules formed in the transition from the alpha to beta forms, and vice versa. Recall from Chapter 5 that a chemical equilibrium is a dynamic one. Molecules of glucose are constantly going from alpha-ring form to straight-chain form to beta-ring form and back again to straight-chain and alpha-form. Thus there are, in fact, always some open-chain molecules present. As long as these open-chain forms are "removed" by being oxidized by Fehling's solution, the equilibrium will continue to be upset. Thus both the alpha and the beta forms will continue to form straight-chain molecules until all of the glucose has been oxidized or all of the test solution has been reduced.

8–3 CHEMICAL CHARACTERISTICS OF CARBOHYDRATES

Carbohydrates may differ from each other in two primary respects, both of which are of great importance to living organisms. First and foremost is their solubility in water. No substance can be biologically utilized unless it is first made soluble. This, essentially, is the job of digestion.

A second factor is one of molecular size. A large molecule, even if soluble, does not normally pass by diffusion through the intestinal

---

* The use of Fehling's solution in this manner is often called a test for the presence of simple sugars. This is not precisely correct, since Fehling's solution will give a positive reaction to any *reducing* sugar, i.e. a sugar molecule with an aldehyde or ketone group exposed. A positive test will *not* be given to the ring structure of a simple sugar molecule, while a positive test *will* be given to a double sugar with an exposed carbonyl group.

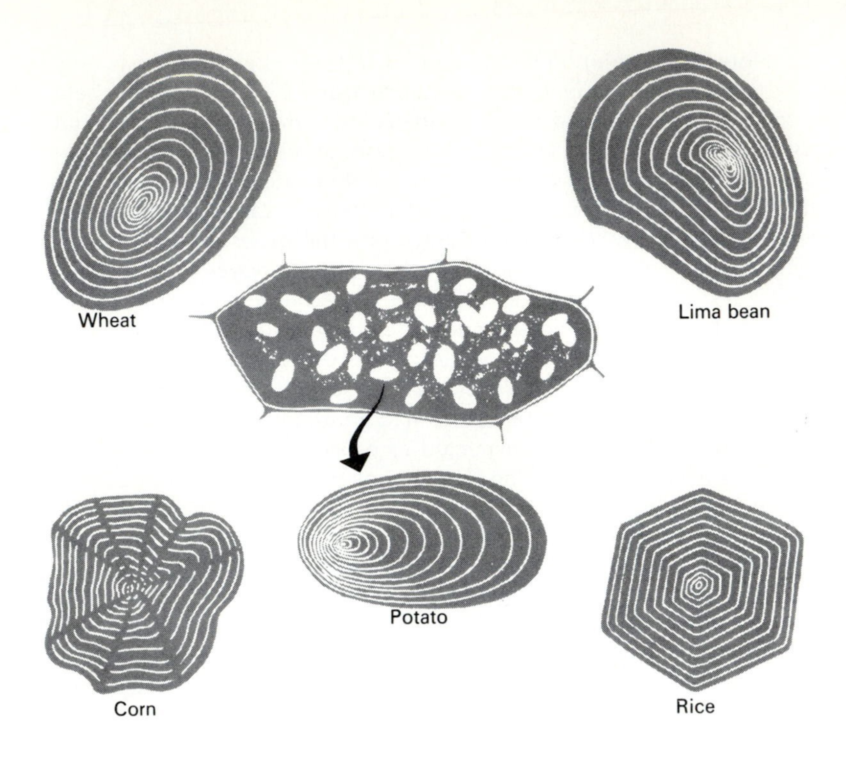

8–2
Starch grains are found scattered within the cytoplasm of plant cells. They can easily be seen under an ordinary light microscope. Note that the starch in each grain is deposited in rather concentric layers. (Adapted from James Bonner and Arthur W. Galston, *Principles of Plant Physiology*. San Francisco: W. H. Freeman and Co. Copyright © 1952.)

surface into the bloodstream. Nor does it permeate cell membranes. Soluble milk sugar(lactose, a disaccharide), for example, is still broken down into smaller molecular parts (monosaccharides) to be absorbed.

On the other hand, there are definite advantages to having carbohydrates in living matter which are either insoluble or which have molecules too large to diffuse through cell membranes. Such molecules are easily stored. For example, *starch* is found in plants in the form of somewhat spherical structures called *starch grains* (Fig. 8–2). The inner portion of each grain is the soluble starch, *amylose*. The grain's outer layer is composed of the relatively insoluble starch, *amylopectin* (Fig. 8–3). Both amylose and amylopectin can be split into the soluble glucose molecules of which they are composed. These glucose molecules can then pass out of the cell and be transported to wherever they are needed.

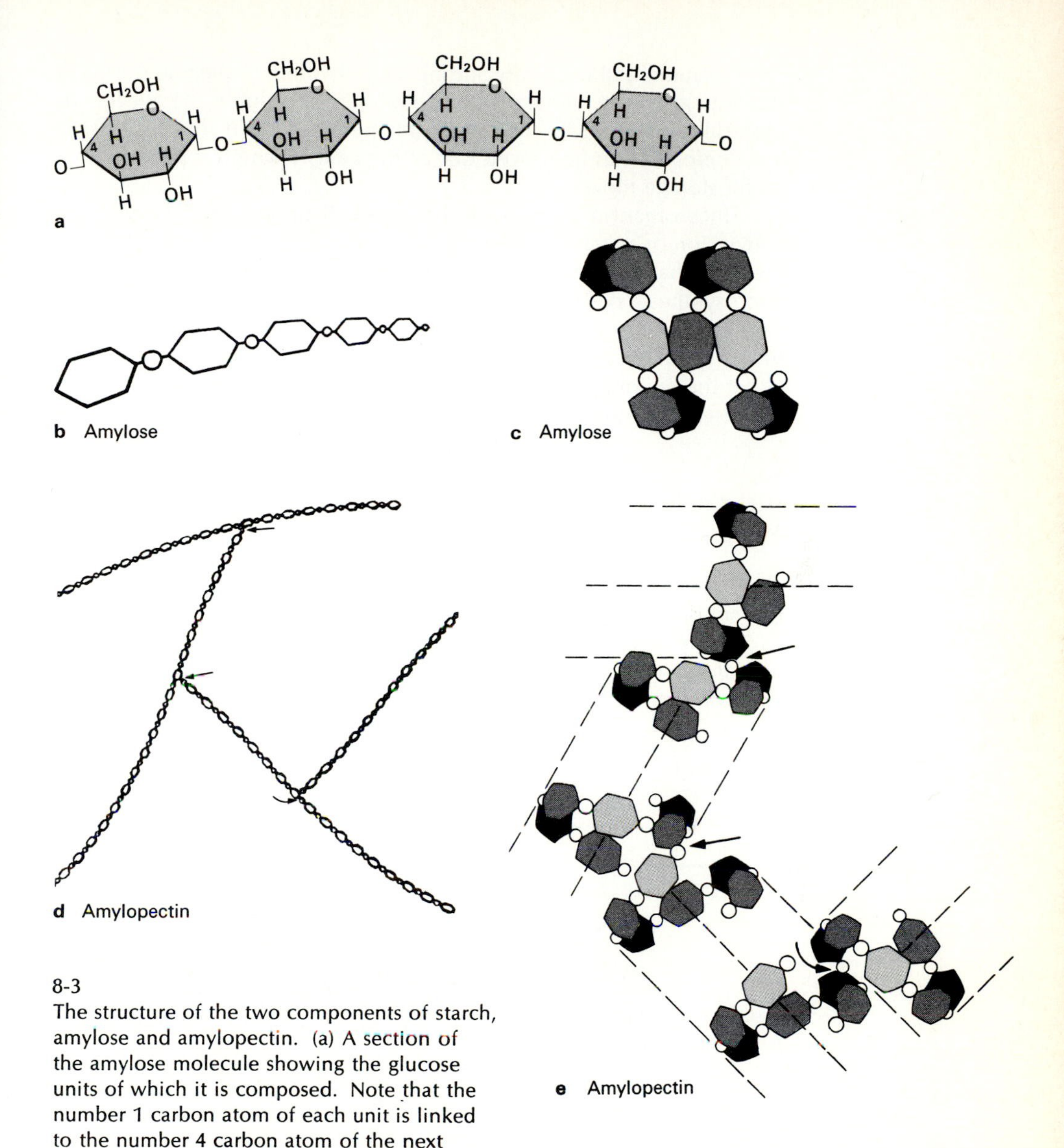

8-3
The structure of the two components of starch, amylose and amylopectin. (a) A section of the amylose molecule showing the glucose units of which it is composed. Note that the number 1 carbon atom of each unit is linked to the number 4 carbon atom of the next unit. (b) An uncoiled representation of the same amylose chain. Each link represents one glucose unit. (c) The chains of glucose units are twisted into a spiral with 6 glucose units to each turn. (d) Uncoiled amylopectin is similar to amylose except that the chains of glucose units branch off each other with each branch averaging about 15 units. Note that where the branching chains are joined to each other (arrows) a different linkage occurs than the 1,4 linkage which exists between the straight-chain units. (e) The branched chains of amylopectin are also twisted into a spiral. The arrows point to the same linkages as those marked in (d). The significance of these different linkages becomes apparent in the digestion of starch (Sections 8–4 and 8–5). Adapted from James Bonner and Arthur W. Galston, *Principles of Plant Physiology*. San Francisco: W. H. Freeman and Co. Copyright © 1952.)

In animals, the corresponding carbohydrate molecule to plant starch is *glycogen.* Unlike starch, glycogen is soluble. However, its molecular size is such that it must be chemically split into its glucose units before it can be moved out of the cell. Thus glycogen is also a useful storage molecule.

Since organisms frequently have to call upon food reserves, the conversion of starch or glycogen to glucose goes on all the time. On the other hand, glucose is constantly being produced by the plant in photosynthesis or being taken into an animal with its food. If not needed immediately, the glucose must be stored. Thus a great deal of the metabolism of carbohydrates within living organisms concerns either the building of large molecules for storage or the tearing down of these large molecules for immediate use.

8–4 THE FORMATION OF DISACCHARIDES AND POLYSACCHARIDES

As Table 8–1 indicates, the basic carbohydrate structural units are the simple sugars, or *monosaccharides.* These are the bricks with which the larger carbohydrates are constructed. For example, the joining of two glucose molecules yields maltose, a double sugar, or *disaccharide* (Fig. 8–4a):

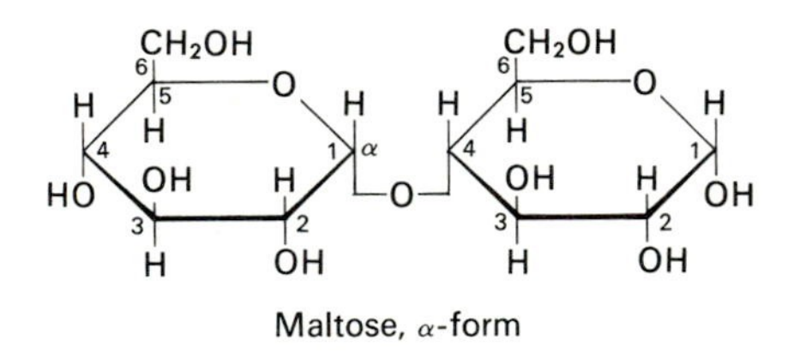

Similarly, the uniting of galactose and glucose yields the disaccharide, *β*-lactose (Fig. 8–4b)

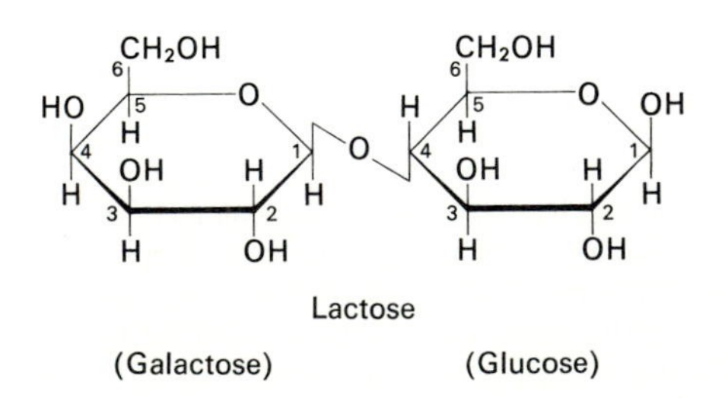

Glucose and fructose can be united to yield ordinary table sugar, sucrose:

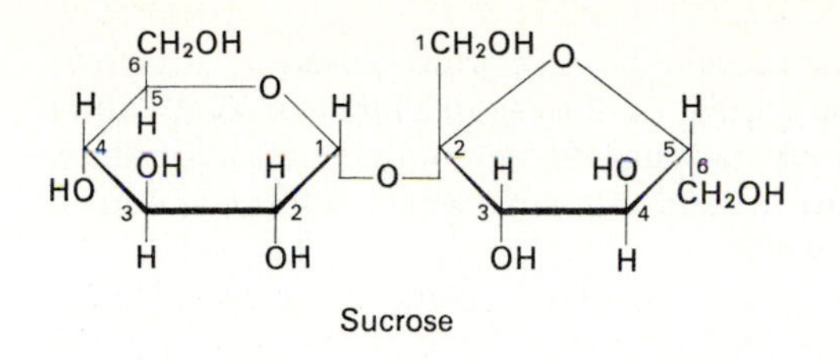

But the disaccharides are only the beginning in the construction of complex carbohydrates. Three monosaccharides joined together yield a *trisaccharide,* such as raffinose. The number of monosaccharides that can be joined is almost limitless.

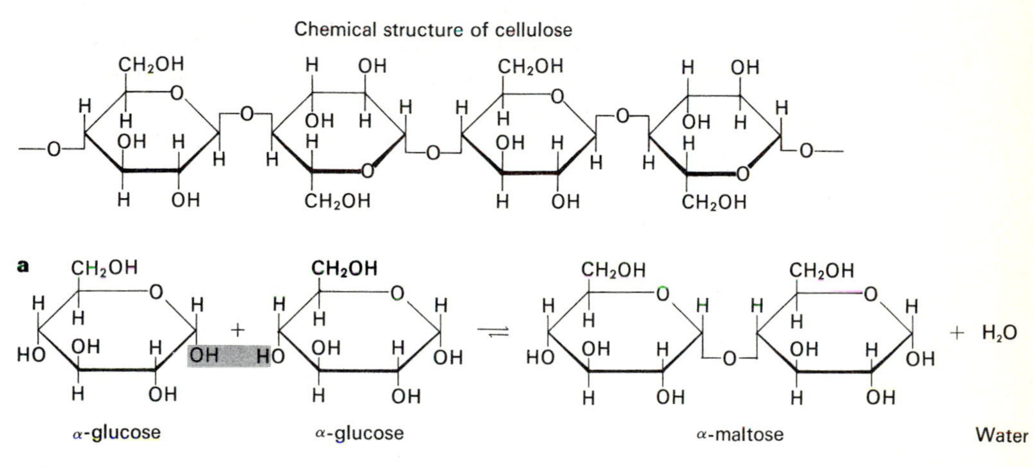

8–4
Alpha (α) and beta (β) linkages. (a) Alpha linkage is formed when, for example, two α-glucose molecules are linked together by splitting out a water molecule, forming an α-maltose molecule and a water molecule, as illustrated above. (b) Beta linkage is formed when, for example, a β-galactose molecule and an α-glucose molecule are linked together by splitting out a water molecule, forming a β-lactose molecule and a water molecule, as illustrated below. (If the righthand ring of the molecule were flipped over, the linkage could be drawn as ┌─O─┐ instead of as ┌─O─┘, although both ways are correct.)

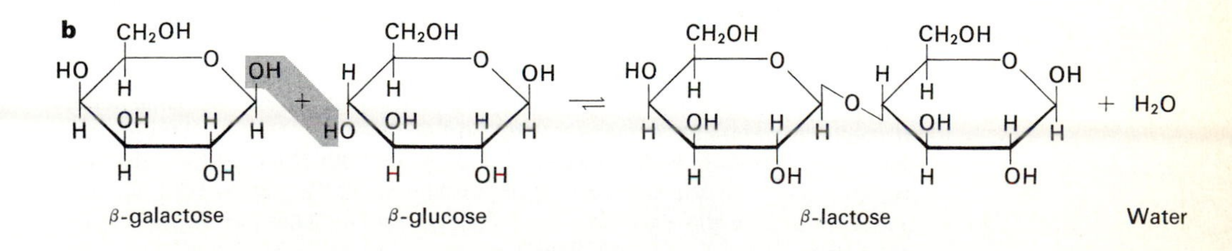

Beyond three units, the compounds formed are generally called *polysaccharides.* The most important polysaccharides to living organisms are *starch, glycogen, chitin* and *cellulose.*

The last two, chitin and cellulose, are widely used as structural material. Chitin forms the bulk of the exoskeleton or hard outer covering of insects and crustaceans (for example, crabs, lobsters). Cellulose is almost entirely limited to plants, where it forms the main constituent of plant cell walls.

However, as nutrient substances, cellulose and chitin are of lesser importance. Many animals, including man, have no enzymes capable of digesting them. In plant-eating animals (herbivores), certain microorganisms in the digestive tract which possess the proper enzymes (yeasts, protozoa, or bacteria) attack cellulose to form glucose. The glucose can be used by the host animal for energy purposes. Termites use cellulose-digesting intestinal protozoa to aid them in their breakdown of wood. Wood, being mostly plant cell wall material, is constructed largely of cellulose.

Starch is man's primary source of carbohydrate food. It is stored by plants, usually in the stems (potato), roots (carrot) or seeds (wheat). Glycogen, often called "animal starch," is stored in the liver and muscles. The quick transport of digested and absorbed foods (via the portal vein running directly from the intestines to the liver) enables the quick transformation of glucose and other monosaccharides to glycogen. As is the case with starch in plants, this glycogen can be broken down as the need arises into usable glucose molecules.

8–5 CARBOHYDRATES AS FUELS

Carbohydrates serve as the primary chemical energy source for almost every form of life. The oxidation of carbohydrates is a vital chemical function of the plant or animal body.

It is certain that no living organism can exist without carrying out at least some form of oxidation. This oxidation can be represented as:

FUEL + OXIDIZING AGENT ⟶ FUEL FRAGMENTS + ENERGY

In most organisms, the fuel is the monosaccharide, glucose. This sugar, therefore, is the nutrient upon which all other life processes depend. Hence we can rewrite the above representation chemically as:

$$\underset{\text{(Glucose)}}{C_6H_{12}O_6} + 6O_2 + 6H_2O \rightarrow 6CO_2 + 12H_2O + \text{ENERGY}^*$$

* The equation shown here differs from the equations for the same series of reactions given above in Eqs. (4–19) and (5–2) and (5–4). The difference is that in the above equation, water is shown on *both* sides. Why not strike out the $6H_2O$ on the left-hand side, balancing the equation by including only 6 waters on the right? Water molecules actually take part in the breakdown process of glucose, and thus are active components of the reaction. The earlier form of the equation which we have used gives only a balance sheet. The longer form indicates more clearly what the reactants in the oxidative process are.

This is one form of respiration. By this oxidative process, a great deal of potential chemical energy is released and stored for immediate or future use in the organism. The equation above is purely representative and shows only the raw materials and end products. The entire respiratory process is far more complex.

Studies made on starving animals indicate that the body uses carbohydrates first, fats second, and proteins last as sources of energy. This is directly proportional to the ease and efficiency with which these substances can be used. Since the carbohydrates are the main source of energy for most organisms, they usually comprise 60–80 percent of a balanced diet.

The carbohydrates need not always remain as such within the body of a living organism. By removing some of the carbon-12 atoms of the carbohydrate molecule and replacing them with carbon-14 (a radioactive isotope), it can be shown that carbohydrates taken into the body are often converted into fat. The fat may then become incorporated into structural components of the living cell. This serves to emphasize again the dynamic, changing nature of living material.

PART 2
LIPIDS

8–6 GENERAL CHARACTERISTICS OF LIPIDS

The lipids are a group of organic compounds that include the fats (nonpolar molecules) and the complex lipids (polar molecules). The fats are important food storage molecules. The complex lipids serve important structural roles in the cell membranes. All lipid molecules, like the carbohydrates, contain the elements carbon, hydrogen, and oxygen. However, unlike the carbohydrates, the ratio of hydrogen to oxygen is far greater than 2:1. The complex lipids also contain phosphorus and often nitrogen. The significance of these facts will be discussed shortly.

The true fat molecule has two parts. These are (1) an *alcohol* (usually glycerol*) and (2) a group of compounds known as *fatty acids*. Fats are broken down into these two parts during digestion.

---

* Also called *glycerine*, or simply, *glycerin*.

TABLE 8–2

Formula	Acid name	Commonly found in
CH_3CH_2COOH	Butyric	Butter
$CH_3(CH_2)_{14}COOH$	Palmitic	Animal and vegetable fats
$CH_3(CH_2)_{16}COOH$	Stearic	Animal and vegetable fats
$CH_3(CH_2)_{24}COOH$	Cerotic	Beeswax, wool fat, etc.

The glycerol portion of the molecule has the structural formula shown below.

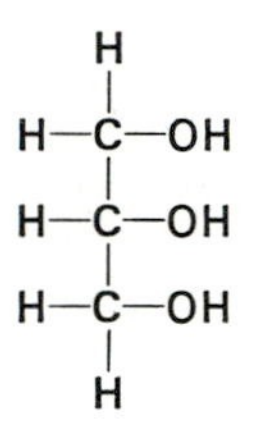

The shaded portion, the alcohol groups, indicates the region where fatty acids can be attached.

There are many kinds of fatty acids, which differ mostly in molecular size. A few of them are listed in Table 8–2.

It can be seen that all of these consist of a *carboxyl* group (—COOH), attached to varying numbers of carbon and hydrogen atoms. The carboxyl group is found in all organic acids. The joining of a fatty acid to the glycerol molecule is accomplished by the removal of an $—H^+$ from the glycerol molecule and an $—OH^-$ group from the fatty acid. The H^+ and OH^- unite to form water. Since glycerol has three —OH groups available, three fatty acid molecules can be attached, as shown below:

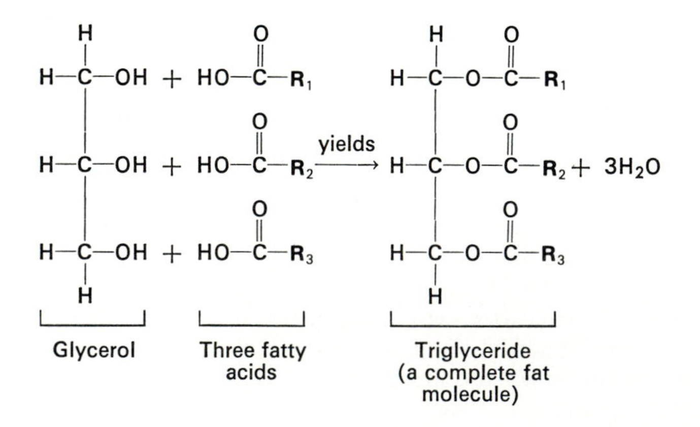

The linkage group between glycerol and fatty acid molecules is the *ester* group. Three ester groups are shown below at the union point between the glycerol and the fatty acids.

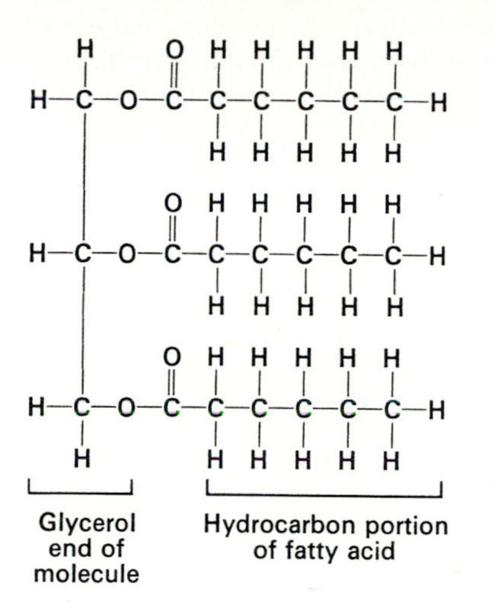

The splitting of fats involves breaking the ester linkage between glycerol and fatty acids. The enzyme *lipase,* aided in its work by emulsification, catalyzes this reaction. Glycerol, as well as short-chain fatty acids of ten or fewer carbon atoms, is absorbed directly into the bloodstream. Fatty acids longer than ten carbon atoms enter the lymphatic system. Eventually, they pass into the bloodstream through the thoracic duct, which opens into the left subclavian vein above the heart.

Some undigested fats may be absorbed directly into the lymphatic system, if the molecules are dispersed finely enough. Such a fine dispersion does occur in a mixture of bile, free fatty acids, and monoglycerides.* It has been estimated that as much as two-thirds to three-quarters of ingested fats are absorbed in their original undigested state.

As previously indicated, the R-portion of the fatty acid molecule is the part to which the carboxyl group remains attached after the lipid molecule has been split. Note that in the list of fatty acids, Table 8–2, only two elements, carbon and hydrogen, are found in the R-portion of the molecule. For this reason, it is called the *hydrocarbon* end.

---

* The function of bile, which is released by the gall bladder, is to cause emulsification. In an emulsified state, a larger molecular surface area is produced for attack by lipase. Bile also aids in making the water-insoluble fatty acids more soluble.

If the structural formulas for the fatty acids given in Table 8–2 were drawn, it could be seen that no more elements can be added to the carbon chain. All of the bonds are filled with hydrogen atoms. However, only *saturated* fatty acids are given in the table. No unsaturated fatty acids are listed. The basis of this distinction can be seen by comparing the complete structural formulas of saturated and unsaturated fatty acids (Fig. 8–5).

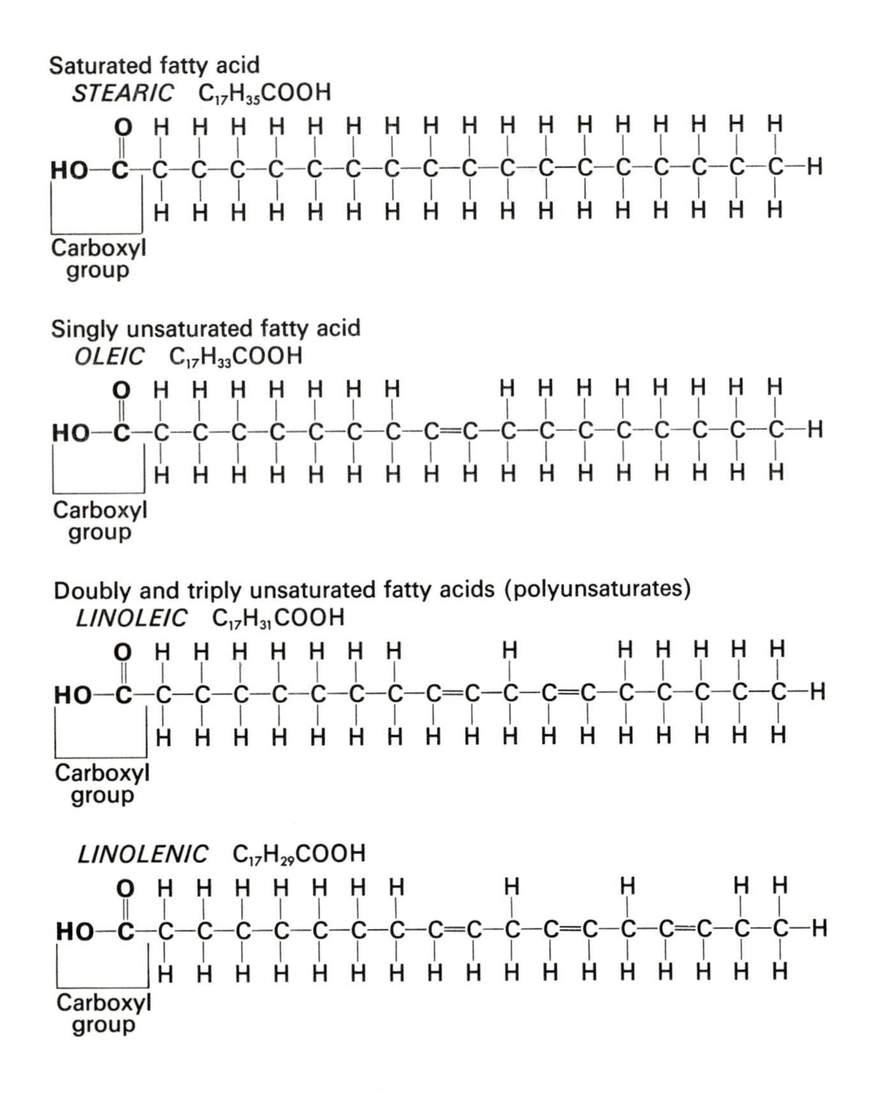

8–5
A comparison of four fatty acids showing the distinction between the saturated and the unsaturated states.

The degree of unsaturation is directly related to the number of double bonds occurring between the carbon atoms. The term "saturation" comes from the fact that all of the carbon atoms have their four electrons shared with different hydrogen atoms. Thus no more may be added to the compound without introducing another carbon atom. A singly unsaturated fat has a hydrocarbon chain with one double bond. It can therefore take on two additional atoms, such as those of hydrogen. It would then become saturated. Similarly, doubly and triply unsaturated fats would become saturated by taking on 4 or 6 atoms, respectively. In general, the melting point of a fat is inversely related to the degree of unsaturation of its fatty acids. The higher the degree of unsaturation, the lower the melting point. When fats melt we call them "oils."

Over the past few years, the presence of saturated fatty acids in the human diet has been related to the occurrence of cardiovascular disease, principally to atherosclerosis (hardening of the arteries). The physiological mechanism of this process is complex, and the actual relationship of saturated and unsaturated fatty acids in the diet to the incidence of atherosclerosis is not completely understood. However, the scheme, as medical researchers now understand it, provides a good example of the interacting chemical systems involved in the body's functions. It appears that the clogging of arteries and veins—from gradual build-up of deposits along the inner walls or from the formation of clots which pull loose and obstruct the central passageway—is directly related to the amount of blood cholesterol. Cholesterol is a steroid, a special group of molecules synthesized from the breakdown products of fats, and forming the basis of many of the body's hormones. The more blood cholesterol, the greater the rate of deposition of substances on the inner walls of arteries. Dietary cholesterol, on the other hand, does not appear to increase the rate of deposition, since cholesterol ingested orally is broken down by the digestive enzymes and thus does not enter the blood stream as cholesterol. Most of the body's cholesterol is synthesized in the liver, through a complex chemical reaction which has an equilibrium point shifted to the left (away from the formation of cholesterol). Normally, the body synthesizes only small amounts of cholesterol. However, if a person eats foods rich in saturated fatty acids, the equilibrium of the cholesterol-forming reaction is disturbed, and more cholesterol is synthesized. Over a long period of time this can lead to serious blockage of the vascular system.

Now, some researchers were pointing out not too long ago that eggs are rich in cholesterol and hence, if eaten in any quantity, could lead to atherosclerosis. In general, most foods which contain high cholesterol levels also contain high levels of saturated fatty acids. Eggs are high in cholesterol but low in saturated fatty acids, and hence appear to be much safer to eat than formerly suspected.

Fats are an organism's most concentrated source of biologically usable energy. Most of them provide twice as many calories per gram as do carbohydrates, though the latter are known as the "energy foods." However, the efficiency of fats as metabolic fuels is 10–12 percent less than that of carbohydrates.

The chemical reason that fats contain more energy than carbohydrates should be evident from a comparison of molecular formulas. $C_{57}H_{110}O_6$ is a fat. $C_6H_{12}O_6$ is the carbohydrate fuel, glucose. Energy is released in these compounds by *oxidation*. The greater hydrogen content of fats means that they are capable of a greater degree of oxidation. Thus they can supply more energy.

Not only do fats supply more energy, they also release a great deal of water when broken down. This gives a double reason for animals to store fats prior to hibernation. The fats supply a great deal of energy per gram, and also enable the animal to survive without drinking water.

The complex lipids mentioned earlier are located primarily in biological membranes, where they play important structural roles. The complex lipids are often called phospholipids because a phosphate group is always part of the molecule. A phospholipid contains glycerol and two fatty acids attached to it by ester linkages. The third hydroxyl group (OH) of the glycerol is attached to a phosphate group. This phosphate group may have an additional group attached to it. The phospholipid shown below contains choline attached to the phosphate.

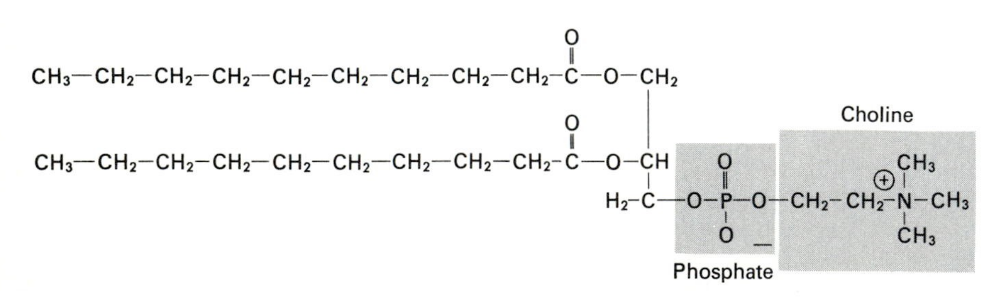

There are several things to notice about this molecule. The phosphate group linked in this fashion bears a negative charge on an oxygen atom. The choline portion contains a nitrogen atom covalently bonded to four other groups. This results in its bearing a positive charge, as does the ammonium ion, NH_4. (Consult Eq. 6–15 on page 96.) This particular phospholipid is called phosphatidylcholine. In other phospholipids the choline is replaced by serine, ethanolamine, or glycerol.

Although as a whole this molecule is electrically neutral, the location of charges at specific sites on the molecule will result in the attraction of other charged or polar molecules to these sites. For example, water molecules will cluster around the charged part of the phosphatidyl choline, but will have no affinity for the fatty-acid hydrocarbon chains. The consequence of this is that the phospholipids will orient themselves on a water surface (see Fig. 3–7, page 45).

The hydrocarbon chains of the fatty acids have no affinity for water (they are said to be "hydrophobic"). They do however have an affinity for one another. The consequence of this is that phospholipids do not dissolve in water but form aggregations, so that the charged portions of the molecules (the "hydrophilic" portions) interact with water, but the hydrocarbon chains interact only with other nonpolar hydrocarbons. Such aggregations of phospholipids have specific forms when dispersed in water (Fig. 8–6(a), (b)).

Similar principles are involved in orienting the lipid molecules to each other, and to the protein components of biological membranes. In 1972 Singer and Nicolson proposed the fluid mosaic model

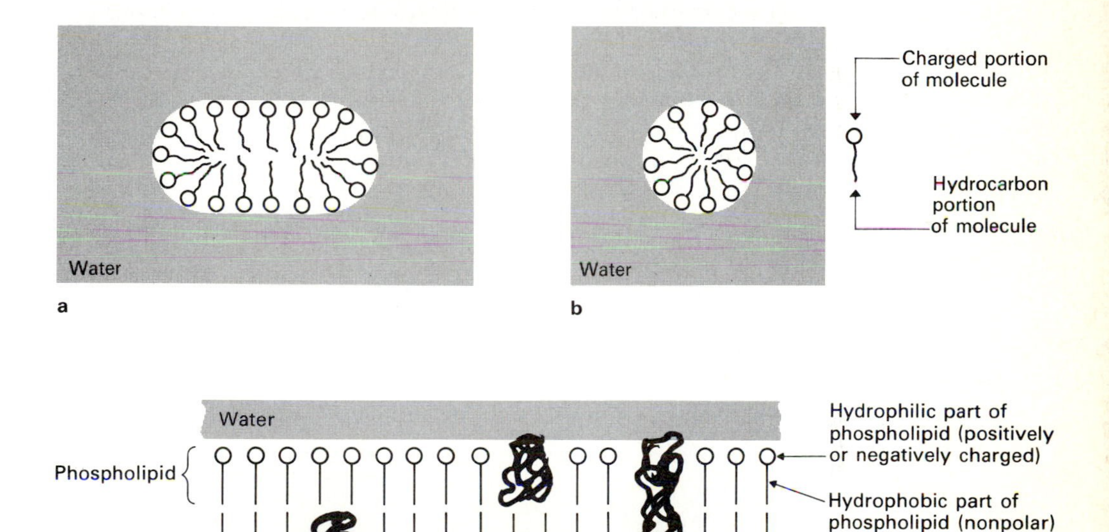

8-6
(a) A lamellar (leaflike) aggregation of phospholipids in water. (b) A micellar (spherical) aggregation of phospholipids in water. In both these cases the charged portions of the molecules are on the outside of the aggregation, in contact with the polar water molecules. The hydrocarbon chains of the fatty acids are in contact with each other in a region which contains no water. (c) The fluid mosaic model of the structure of biological membranes. The charged portion of the phospholipid may bear a positive or negative charge, depending on the specific type of phospholipid. The hydrocarbon chains have no affinity for water or for the charged protein, and associate together in the center of the membrane.

of membrane structure. This model proposes a phospholipid bilayer with protein interspersed, thus permitting the proteins to form channels for transport through the hydrophobic portion of the membrane (Fig. 8–6(c)). This general structure is characteristic of the membranes of many living cells, from bacteria to those of higher animals and plants.

8–7 CARBOHYDRATES AND FATS: CONSTRUCTION AND DESTRUCTION

The chemistry of life is never still. Thousands of molecules are being broken down and put together every second. It is this constant chemical activity which is referred to whenever living organisms are spoken of as being "dynamic systems."

Carbohydrates and fats are very much involved in this activity, since both are important sources of energy. For example, starch and glycogen are constantly being broken down into glucose for use as fuel. Elsewhere, glucose molecules are being united to form starch or glycogen as a means of storing the glucose for future use. Fats are

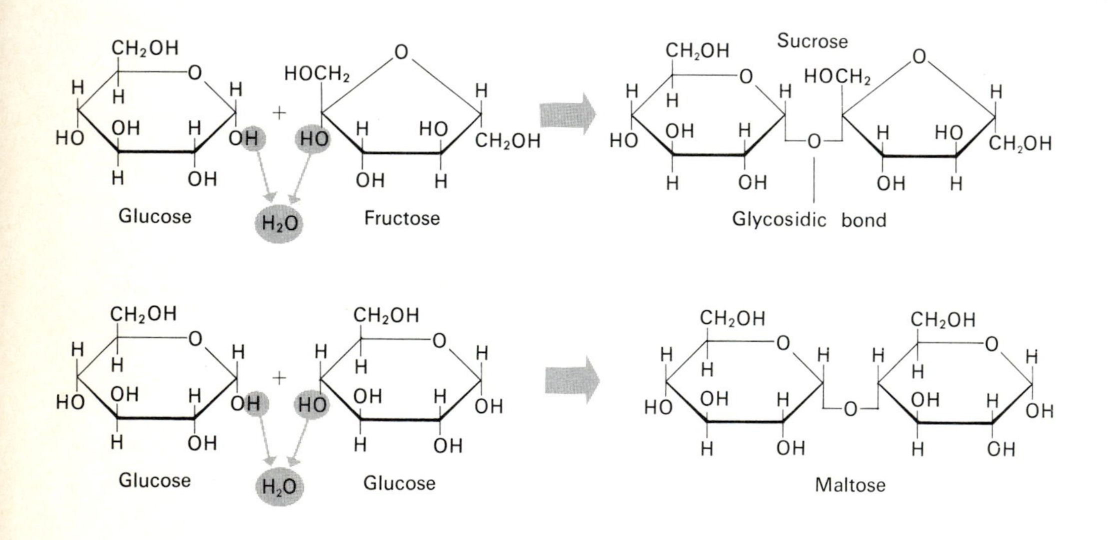

8-7

Formation of two disaccharides by dehydration synthesis. Top, a glucose and fructose molecule form sucrose (table sugar); bottom, two glucose molecules form maltose. The removal of a water molecule between the two molecules of the simple sugars forms a glycosidic bond, of which there are several types, depending on the simple sugars involved.

split to form fatty acids and glycerol. At the same time, fatty acids and glycerol are being chemically united to form fats.

For these destructive and constructive reactions, living organisms use the same chemical techniques. The union of monosaccharides to form disaccharides takes place as shown in Fig. 8–7.

The union of glycerol and three fatty acids gives a fat molecule, as follows:

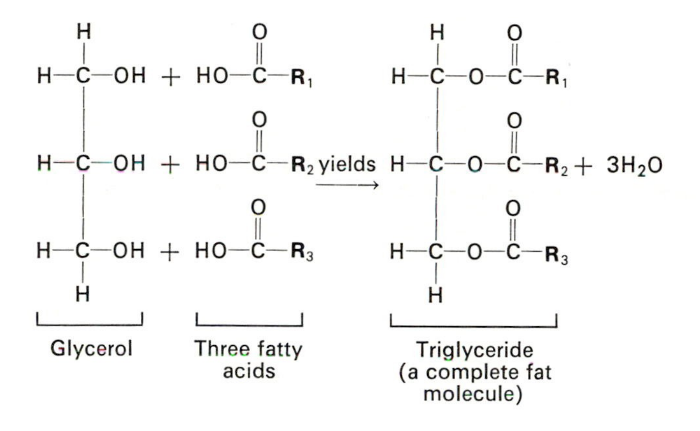

Note that in each case the constructive reactions, or syntheses, involved the removal of one or more water molecules. These reactions, therefore, are called *dehydration synthesis* reactions. A disaccharide becomes a polysaccharide as more monosaccharide units are added. At each addition, a water molecule is removed. The polysaccharides starch and glycogen are formed by a long series of dehydration syntheses.

Just as the removal of water molecules results in synthesis, so the chemical addition of water molecules results in breakdown. Reactions in which the chemical addition of water results in the splitting of molecules are called *hydrolysis* (*hydro-*, water; *lysis*, breakdown) reactions. The digestion of foods proceeds by hydrolysis. This is one of the reasons why water is essential for digestion to occur.

The polysaccharide starch is split into glucose molecules in several steps. This can be shown in generalized form as:

STARCH (insoluble and too large)

$+H_2O$ ↓

DISACCHARIDE (soluble but still too large)

$+H_2O$ ↓

MONOSACCHARIDE (soluble and small)

Fats are also split into fatty acids and glycerol by the addition of water molecules at each linkage. Hydrolysis of fats to fatty acids and glycerol is called *saponification*. It can be carried out enzymatically in the living system, or by heating in strong alkali, such as sodium hydroxide (lye). The former process occurs in the digestive system (gastrointestinal tract) of animals, breaking the large fat molecules down so they can be absorbed into the circulatory system. The latter process is the way soap is produced from animal fats.

Dehydration synthesis and hydrolysis are not limited to carbohydrates and fats. Many other compounds, including proteins, are synthesized and broken down by dehydration synthesis and hydrolysis, respectively.

THE UPTAKE OF GLUCOSE—GLUCOSE TOLERANCE TEST

It has already been seen how glucose concentration can be measured (see page 87). In certain cases a physician may wish to determine not only the amount of glucose in the blood after a 12-hour fast, but also the rate of glucose uptake from the bloodstream into the tissues. The rate of uptake may be an indicator of the ability to use glucose as a fuel.

A glucose solution is administered orally to a patient who has fasted at least 12 hours. At that time, and every hour afterward for six hours, a blood sample is drawn. Blood glucose is measured as described previously (see page 87) and a graph is constructed plotting mg glucose/100 ml (vertical axis) *vs.* time in hours (horizontal axis) (Fig. 8–8). The physician may make a diagnosis based on the shape and elevation of the curve.

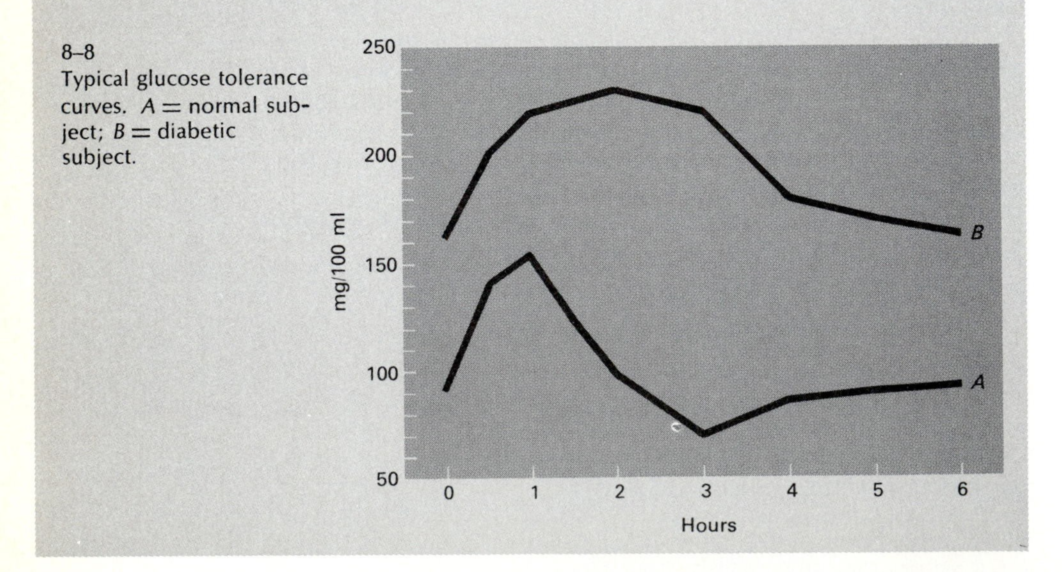

8–8
Typical glucose tolerance curves. $A =$ normal subject; $B =$ diabetic subject.

In a normal patient the concentration of blood glucose will be high for approximately one hour after the administration of the glucose solution. The high blood-glucose concentration acts as a stimulus for the pancreas to release a large amount of insulin into the bloodstream. (Insulin is a hormone that causes muscle and liver cells to take up glucose from the bloodstream and to utilize glucose more rapidly.) This causes the blood-glucose concentration to fall rapidly between the second and third hours of the test. Once the blood-glucose concentartion is low, the secretion of such a large quantity of insulin is no longer stimulated and blood glucose levels off to a fairly constant level for the remainder of the test (hours 4, 5, and 6).

In a certain type of diabetes, insulin production or secretion is deficient. This results in relatively high blood-glucose concentrations even after fasting (160 mg/100 ml in the diabetic *vs.* 90 mg/100 ml in the normal individual), and a much higher blood-glucose concentration one hour after administration of the glucose solution. In addition, the utilization of glucose is not as rapid, so that the blood-glucose concentration will remain high for approximately three hours before a gradual decline is observed. (This is in contrast to the rapid decline observed at two hours in the normal patient.) Thus it can be seen that, in the diabetic individual (*B*) in whom insulin production/secretion is deficient, the utilization of glucose as a fuel is less efficient than in the normal individual (*A*). See Fig. 8–8.

Source: White, Handler & Smith, *Principles of Biochemistry* (Fifth ed.), McGraw-Hill, New York, 1973, p. 845 and Fig. 18.8.

ALCOHOL: ITS METABOLISM AND SOME OF ITS EFFECTS

The alcohol found in alcoholic beverages is ethyl alcohol. This is one of the few substances that can be absorbed into the bloodstream directly through the stomach and even more rapidly through the intestine, which has a larger surface area. As much as two-fifths of the alcohol is absorbed directly through the stomach wall. The kind of drink, whether it is diluted, and whether there is food in the stomach are all factors that affect the rate of absorption into the blood. If the consumption of alcoholic beverages is preceded by the intake of milk or a cheese-dip, the rate of emptying from the stomach is slowed down.

The distribution of alcohol through the body is not proportional. Alcohol is maintained in a relatively high concentration in the blood and cerebrospinal fluid, is absorbed to a lesser extent in muscle tissue, and hardly at all into bone and fat tissue. The solubility of alcohol in water may help to explain why alcohol is distributed principally to those parts of the body that have a high water content. Since the brain is interlaced with blood vessels and surrounded by membranes containing cerebrospinal fluid, it takes only a few minutes for the highest concentration of alcohol to be in the brain.

The metabolism of alcohol occurs mainly in the liver. Of three liver enzymes that can oxidize ethanol to acetaldehyde, alcohol dehydrogenase appears to be the most important. It catalyzes the following equilibrium:

$$NAD^+ + \underset{\text{Ethanol}}{CH_3CH_2OH} \rightleftharpoons NADH + H^+ + \underset{\text{Acetaldehyde}}{CH_3CHO}$$

Acetaldehyde is oxidized to acetate (CH_3COO^-) by NAD-requiring aldehyde dehydrogenases. Most of the acetate leaves the liver and is used by other tissues as well as a fuel by entering the pathway of carbohydrate metabolism (by being combined with coenzyme A to form acetyl-CoA).

The liver accounts for more than 70 percent of alcohol metabolism. The excess that is concentrated in the brain depresses the brain's activity. The reticular formation (the part of the brain that funcions in awareness and sleep/wake cycles) is the first to be affected. For this reason reflexes are slowed down. The brain does not register information as quickly as when the person is not under the influence of alcohol. With further increase in alcohol consumption, drunkenness ensues. The function of the cortex (where voluntary movements and thoughts are controlled) is affected, resulting in giddiness, slurred speech, and unsteady walking. The limbic system ("emotion center" of the brain) is no longer inhibited by the cortex and the person may become more emotional than usual.

The wide individual variation in the response to any given dose of alcohol and the relatively quick development of tolerance to alcohol make it difficult to predict with accuracy the kind of behavior that will follow a particular dose. It is possible, however, to predict the probability of a given behavior at different dose levels. For example, the probability of causing an automobile accident is somewhat under one percent when the blood alcohol level is 0.00 to 0.04 percent, but that probability is doubled with a blood alcohol level of 0.06 percent, and sextupled with a level of 0.10 percent.

Sources: Biology Today, CRM Books, 1972, pp. 396, 901–903. Ebert, James D., Ariel G. Loewy, Richard S. Miller, and Howard A. Schneiderman, *Biology,* Holt, Rinehart and Winston, 1973, p. 428; and Fox, Ruth, and Peter Lyon, *Alcoholism, Its Scope, Causes, and Treatment.* Random House, N.Y., 1955, pp. 3–18.

Chapter 9
Proteins

9–1
INTRODUCTION

Proteins play the most varied roles of any molecules in the living organism. As enzymes, they serve to keep all of the various chemical reactions within a cell operating smoothly and continuously. As structural elements, they serve in such places as the contractile fibers of muscle, the spongy supporting tissue between bones, and in hair, nails, and skin.

It is the purpose of this chapter to discuss the chemical nature of proteins and to show what characteristics enable them to play so many roles. We shall also discuss some of the ways in which biochemists have learned about the structure of protein molecules.

9–2
AMINO ACIDS AND THE PRIMARY STRUCTURE OF PROTEINS

The fundamental building block of protein is the *amino acid*. Amino acids are nitrogen-containing compounds with an amino group (NH_2)

and a carboxyl group (COOH). Amino groups give basic properties to amino acids, while carboxyl groups give acidic properties.

A diagram of a generalized amino acid is shown below. Attached to a carbon atom in every amino acid is a characteristic group of atoms symbolized "the R-group."

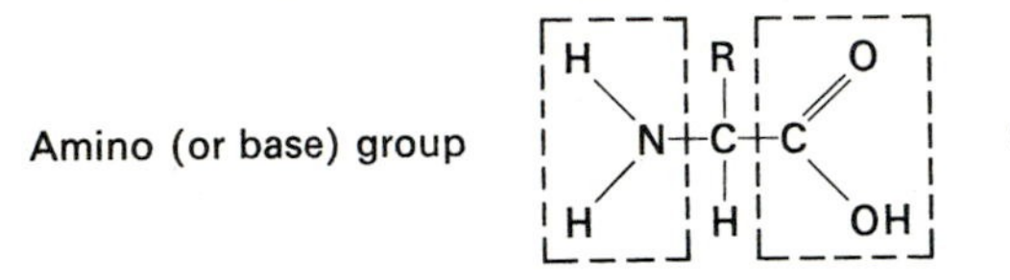

It is in the number and arrangement of atoms comprising the R group that one amino acid differs from another. Otherwise, all amino acids have at least one amino group and one carboxyl group.

From about twenty different kinds of amino acids, all the proteins known to exist in plants and animals are constructed.* Amino acids are to proteins what letters of the alphabet are to words. A group of amino acids can be joined together in a specific order to produce a given protein, just as a group of letters can be arranged to form a word. For this reason, amino acids are often referred to as the "alphabet" of proteins.

Amino acids are linked in end-to-end fashion to form long protein chains. *The variety found among proteins is the result of the types of amino acids composing each and the order or sequence in which these types are arranged.*

The comparison of amino acids to letters of the alphabet is helpful in representing how a small change in a protein molecule can completely change its chemical properties. Change of one letter in a word may cause that word to become meaningless. For example, the word "skunk" conveys one idea, while the word "skank" means nothing at all. Likewise, a change or substitution of one amino acid for another may make an entire protein molecule "meaningless" to the cell. The molecule is no longer able to carry out its function. For example, a certain portion of the human population possess a condition known as sickle-cell anemia. The hemoglobin in afflicted persons differs from normal hemoglobin in only one amino acid out of 300. Such hemoglobin molecules will not combine as readily with oxygen. Persons with this type of hemoglobin generally have a shortened life span. Thus, a change in one amino acid in one type of protein can have far reaching effects on the entire organism.

Sometimes, the removal or addition of one or two letters in a word may change the meaning without making the word senseless. For example, the word "live" can become "liver" or "olive." Sim-

* A table showing the structural formulas for the amino acids occurring in proteins is in the Appendix.

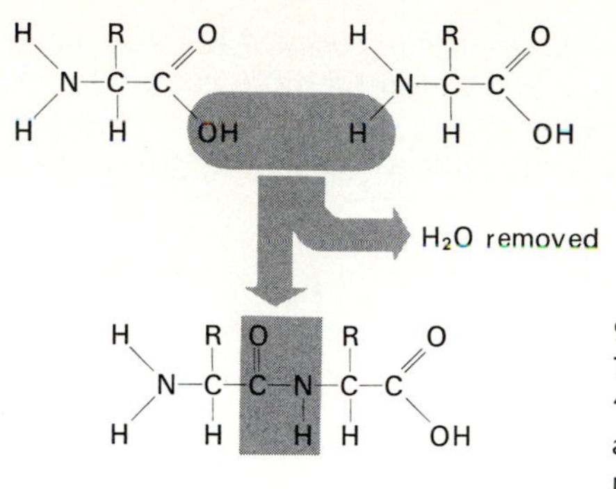

9–1
The formation of a peptide bond, or "linkage," by the joining together of two amino acids. The carboxyl group of one molecule joins the amino group of the other by the loss of a water molecule.

ilarly, one protein may be changed into another by the removal or addition of a few amino acids.

In one way, however, the comparison of proteins to words falls short. This is in the matter of length. Words are generally composed of relatively few letters. Even the name of the New Zealand village Taumatawhakatangihangakoauotamateaturipukakapikimaungahoronukupokaiwhenuakitanatahu only approaches the complexity and length of a small to average protein: Proteins are macromolecules (*macro* = large), often consisting of several hundred to over a thousand amino acids. Their molecular weights range from 6,000 to 2,800,000.

The great size of proteins gives them added versatility in cell chemistry. They can take on a variety of shapes and sizes, each of which may serve very specialized functions. For this reason, biochemists speak of *chemical specificity* as being characteristic of many proteins. It is through the use of such chemically specific proteins as enzymes that living organisms are able to carry out their many different reactions so efficiently.

When amino acids unite to form proteins, the amino end of one amino acid molecule forms a chemical bond with the carboxyl end of the other. The result is the formation of a connecting link between the two, much like the connection of railroad cars. This linkage process continues until all of the amino acids necessary to form the protein are joined together in the order characteristic for that particular molecule. Figure 9–1 shows the joining of amino acids to form this link. Note that this process involves the loss of one water molecule between each two amino acids and is thus a dehydration synthesis. The resulting linkage is called a *peptide bond.*

The joining of amino acids in this manner forms a larger unit called a *peptide.* When thirty or more amino acids are joined together, the entire unit is called a polypeptide. Proteins are polypeptides containing hundreds of amino acids. Some of these large proteins may aggregate into still larger units by lining up end-to-end to form a very large structure. The connective-tissue protein *collagen* is an example. Other proteins aggregate side by side to form a more compact unit. Many large enzymes are examples of this.

The molecular structure of a peptide consisting of four amino acids is shown below:

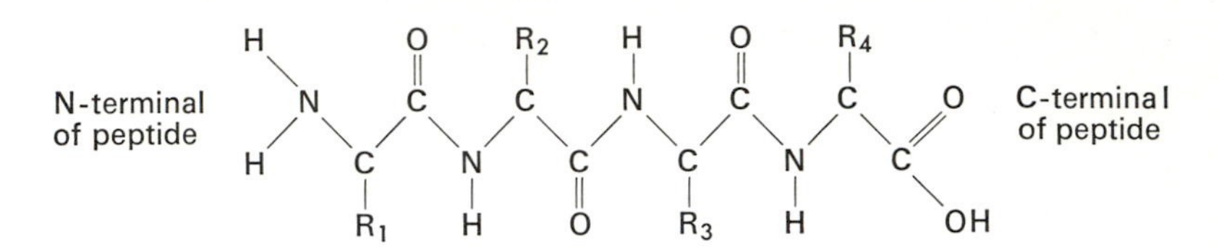

The sequence of amino acids formed from a series of covalent peptide bonds may be considered the most fundamental level of organization of proteins, their "backbone."* This sequence of amino acids is responsible, in part, for the uniqueness of each type of protein. Other factors which contribute to uniqueness of proteins will be discussed in Section 9–3.

The higher levels of protein organization, such as twisting and folding of the polypeptide chain, depend upon the sequence of amino acids for their particular patterns. Why is this sequence of amino acids so important? The R-groups represent those portions of amino acids which make one amino acid different from another. Thus, R_2 may indicate the group of atoms which characterize *lysine,* while R_3 may characterize *alanine* or *leucine*. If one or more amino acids is removed from the backbone, the sequence of R-groups is changed. This can have many effects, but one example will illustrate the point. If two cysteine molecules occur close together in a peptide chain, the sulfhydryl (—SH) groups on each cysteine molecule interact

---

* Currently, the terms *primary, secondary,* and *tertiary* structure of proteins are often used to refer, respectively, to: (1) the amino acid sequence; (2) the coiling of the polypeptide chain into the *alpha helix,* or the interaction of two polypeptides to produce the *beta configuration;* and (3) the folding of the alpha helix into various shapes to produce a more or less globular protein molecule (see following sections). However, the actual meaning of primary, secondary and tertiary structure refers to types of forces stabilizing a protein molecule and not to any actual geometric shape. Thus, primary structure refers to all covalent bonding, mainly peptide and disulfide, within a protein. Secondary structure refers to hydrogen bonding. Tertiary structure refers to electrostatic bonds, interactions between atoms placed extremely close to each other, and the like. In order to avoid discussing the numerous types of bonds that actually contribute to giving a protein molecule its overall stability, we have chosen to discuss only the major geometric patterns which proteins may take. These levels of organization are *not* exactly synonymous with the terms primary, secondary, and tertiary structure.

to form a *disulfide* (S—S) bond. This is represented diagrammatically in Fig. 9–2.

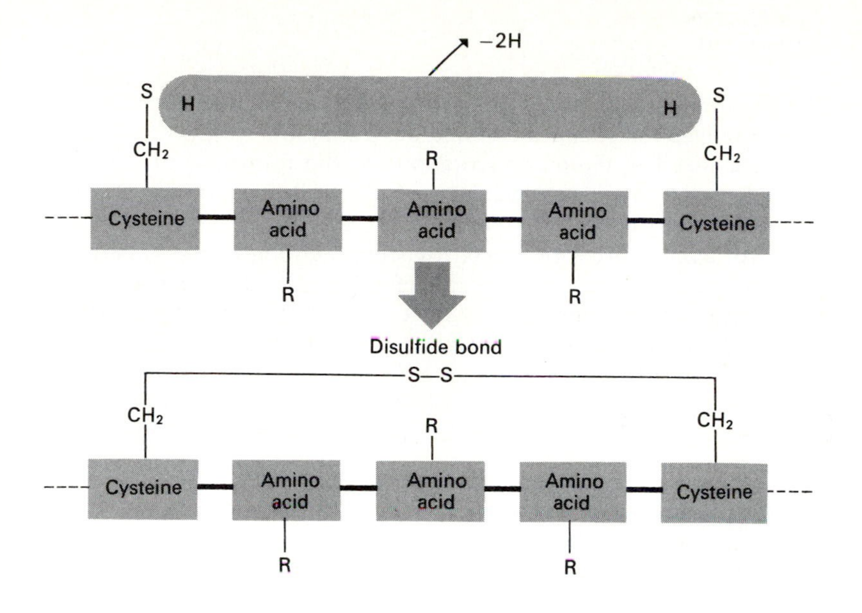

9-2
Formation of disulfide bridge between two cysteines in a protein such as insulin.

Bridges of this sort are important in determining the configuration which polypeptide chains possess. Such bridges may occur between two amino acids of one chain, as shown above, or between two cysteines on separate but adjacent chains. However, such bonds will not form unless the position of one cysteine in respect to that of the other is exactly right. If one amino acid is removed or added between the two cysteines the distance between them is changed. If this happens, no disulfide bonds form. Here is one example of the importance which a specific sequence of amino acids may have to the structure of a protein molecule.

Not all changes in amino acid sequence produce effects of the same magnitude. Removal or substitution of one amino acid for another at one point in the peptide chain may completely destroy the biological activity of a protein. The same change at another point may have no effect whatsoever. Protein molecules apparently have certain sensitive spots where any change in amino acid sequence produces significant changes in chemical activity.

9–3 OTHER FACTORS DETERMINING SHAPE OF PROTEIN MOLECULES

Few proteins exist as a straight-chained sequence of amino acids. Most polypeptide chains are coiled or twisted in a variety of ways and are stabilized in these configurations by the formation of hydrogen bonds between adjacent portions of the molecule. The most prominent configuration is the *alpha helix*.

An alpha helix is produced by spiral twisting of the amino acid chain. To visualize the geometry of a helix, think of a ribbon as a straight-chain polypeptide. A helical structure can be formed by twisting the ribbon several times around a pencil. The spiral which remains after the pencil is removed is the general shape of an alpha helix. Figure 9–3 shows a simplified portion of an alpha helix. The alpha helix is held in position by the formation of hydrogen bonds

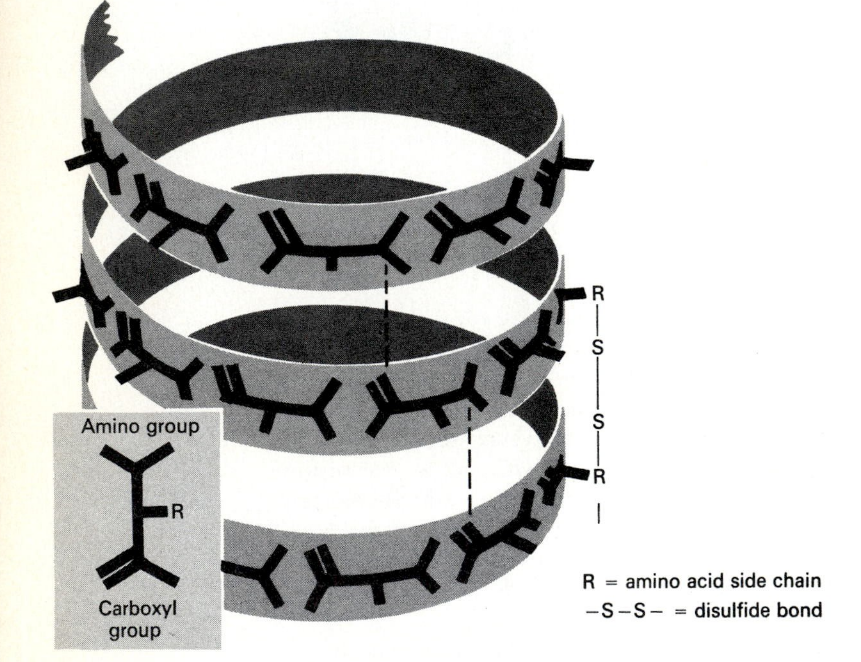

9-3
A diagram representing the alpha-helix formation of protein. The alpha helix (α helix) is one possible configuration of a polypeptide chain. The dotted lines between adjacent turns in the spiral represent hydrogen bonds. Hydrogen bonds serve as the primary forces which hold the alpha helix in its shape. Other bonds which help to give the helix its structure are the disulfide bonds (S—S) and various ionic bonds.

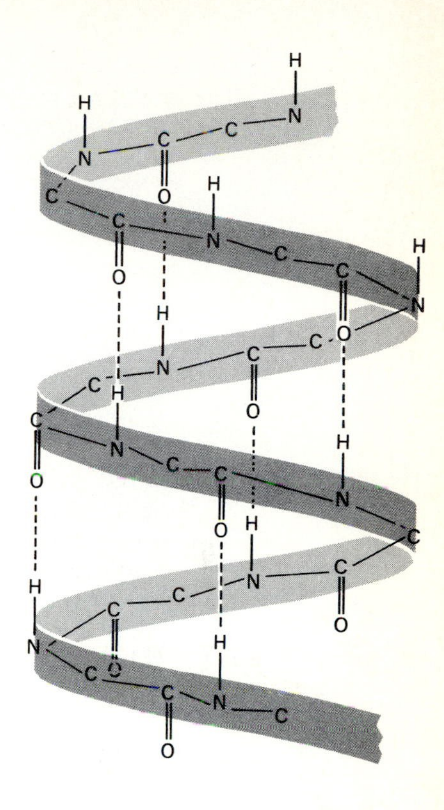

9-4
A diagram of a peptide backbone wound into an alpha-helix configuration. The heavy black curve traces the helical structure. The broken lines going from the C═O to the N—H groups represent the hydrogen bonds which hold the chain in the alpha-helix form.

between amino acids on one part of the chain with those on another part of the same chain.

The hydrogen bonds are formed between the C═O group of one amino acid and an N—H group nearby. Hydrogen bonding is the result of an electrostatic attraction between an unshared electron pair of one atom and the positively charged hydrogen end of a polar molecule. Only a few atoms will form hydrogen bonds, and two of these, oxygen and nitrogen, are found in proteins. Thus, for example, a hydrogen bond can be formed when the two groups shown below approach each other. The dotted line represents the hydrogen bond.

$$\overset{\delta-}{>C \overset{\circ\times}{\underset{\circ\times}{}} \overset{\times\times}{O} \overset{\times}{\underset{\times}{}}}\text{- - - -}\overset{\delta+}{H \overset{\circ}{\underset{\times}{}} N<}$$

It is difficult to overestimate the importance of hydrogen bonding to our present concept of protein structure. Although, individually, hydrogen bonds are quite weak, many hydrogen bonds reinforce each other to produce a stable structure. A more detailed representation of hydrogen bonds in the alpha helix is shown in Fig. 9–4.

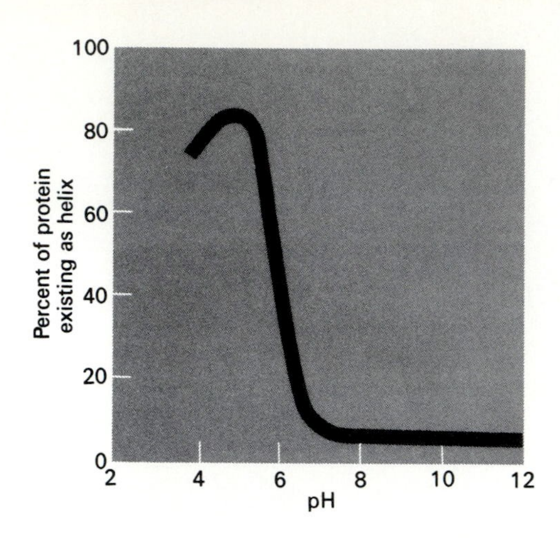

9–5
As the pH of some polypeptide solutions is changed, the amount of alpha-helix formation diminishes suddenly at about pH 5.5. Within a narrow range beyond 5, most of the helical formation is destroyed. The polypeptides then exist as random chains or perhaps as beta configuration (straight chains). (From "Proteins" by Paul Doty. Copyright © 1957 by Scientific American, Inc. All rights reserved.)

Hydrogen bonds can be broken by many physical and chemical means. For example, a change in pH is very effective in weakening hydrogen bonds. When the pH of a protein solution is raised, a sudden change occurs in the number of helical protein molecules. In Fig. 9–5, the sudden plunge which the graph line takes indicates that at pH 5.5 many of the hydrogen bonds break simultaneously. At pH 7.5, virtually all of the hydrogen bonds are broken. Thus a further increase in pH beyond 8 produces no appreciable change. Up to a limit, such a process is reversible. If the pH is lowered to its original value the hydrogen bonds re-form, and the helical structure reappears.

If a protein solution is subjected to gentle heating, similar results are obtained. Heat destroys hydrogen bonds, which allows the helix to uncoil. Upon cooling, the helices reform. If the heating is too vigorous, however, as in boiling, the alpha helix is destroyed. The protein cannot return to its helical structure. The protein is said to be *denatured.* The coagulation of boiled egg white is an example of denaturation. Denaturation is an irreversible process. You cannot "unboil" an egg! Although boiling does not involve a change in amino acid sequence, it changes the spatial orientation of certain groups of atoms enough so that the helix cannot reform. This process is analogous to pulling a coiled spring or "slinky" out of shape (Fig. 9–6). The "slinky" can be stretched for a certain distance and still spring back into shape. If it is pulled too far, however, it cannot resume its former coiled structure. It is then a "denatured slinky."

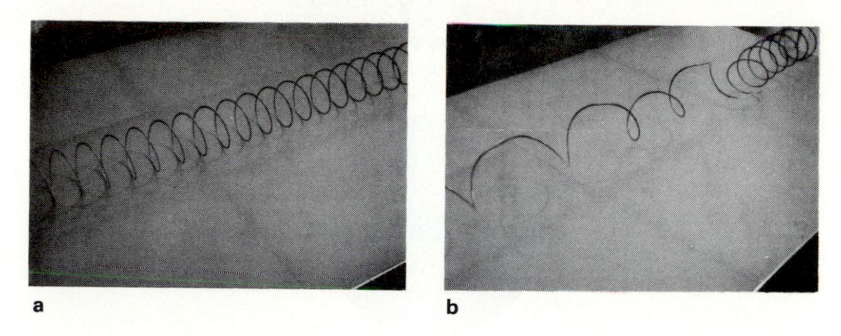

9–6
(a) A coiled spring toy known as a "slinky." This represents the alpha-helix structure of certain proteins. When the slinky is pulled out of shape, as shown in photograph (b), it cannot recoil into the original helical structure. It has been irreversibly altered. This is analogous to denaturing a protein by such physical means as heating or by such chemical means as placing the protein in a concentrated solution of urea.

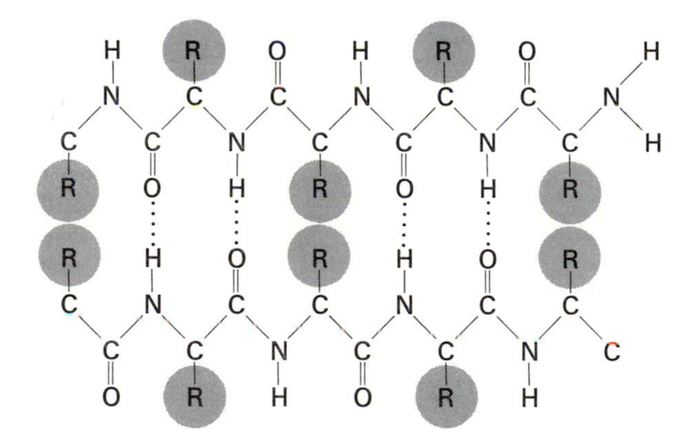

9–7
The beta configuration of a polypeptide chain. In this formation, hydrogen bonds occur between two different peptides or polypeptides, rather than within one chain (as in the alpha helix).

A second type of configuration which polypeptide chains may assume is *beta configuration*. Beta configuration is the result of interaction between two or more polypeptide chains lying side by side. The polypeptide chains in beta configuration are not twisted to form a helix. Beta configuration is found in silk and other protein fibers, as well as in a number of crystalline proteins. Beta configuration proteins, due to their long and unfolded condition, do not form the varied geometrical shapes characteristic of alpha helix proteins. Beta chain configuration is shown in two-dimensional form in Fig. 9–7.

Scale

100 Angstrom units

Hemoglobin
(Molecular weight 68,000)

β_1-Lipoprotein
(Molecular weight 1,300,000)

Fibrinogen
(Molecular weight 400,000)

Glucose molecule on
same scale
(Molecular weight 186)

9–8
The various sizes and molecular configurations for several types of protein found in living systems. Hemoglobin is the oxygen-carrying protein of the blood. It consists of four polypeptide chains. β_1-lipoprotein is found in the liquid portion of blood and in the tissue fluids of higher organisms, such as mammals. It is composed of lipid elements bound to protein and serves in maintaining the stability of certain enzymes. Fibrinogen is also found in the liquid portion of blood and is involved in forming clots. The carbohydrate molecule glucose is included for size comparison.

Proteins with an alpha helix configuration are generally soluble and chemically active. In other words, they take part in the many metabolic reactions within cells. Enzymes, for example, are composed mostly of polypeptide chains in the helical form. Beta configuration proteins are more frequently found as structural parts of organisms, such as bone and cartilage. Chemically they are far less active than the soluble alpha helix proteins.

From what has been said so far, it might appear that soluble proteins are like thin threads or ropes. However, they are actually globular or spherical in shape. The helical polypeptide chain is folded

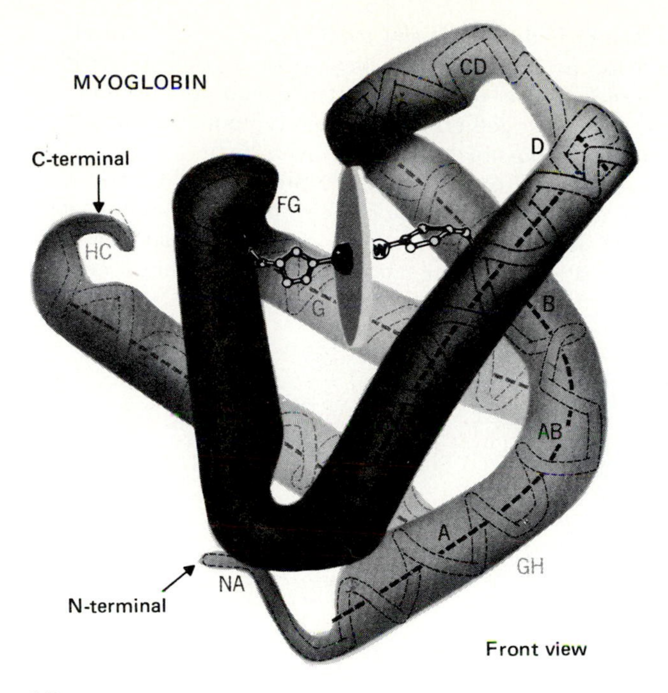

9-9
The myoglobin molecule is built up from eight stretches of alpha helix that form a box for the heme group. Histidines interact with the heme to the left and right, and the oxygen molecule sits at W. Helices E and F form the walls of a box for the heme; B, G, and H are the floor; and the C,D corner closes the open end. The course of the main chain is outlined inside the shaded envelopes. The COOH end of the polypeptide chain is indicated as the C-terminal, the NH_2 end as the N-terminal. Reproduced with permission, from the *Structure and Action of Proteins* by Richard Dickerson and Irving Geis, W. A. Benjamin, Menlo Park, California, © 1969 Richard Dickerson and Irving Geis.

into a compact, rounded molecule. Twisting and folding of the alpha helix gives further chemical specificity to the protein. An extremely important feature of proteins is the fact that the entire molecule has a specific structural form. For each protein, folding of the polypeptide chain is slightly different, giving each molecule its own unique configuration. The shapes of three proteins are shown in Fig. 9–8.

Changes in the folding of an alpha helix can alter the *specificity* of a protein. Specificity of a protein means that each protein will react chemically with only one type of substance. This is thought to be due to the surface configuration of the molecule (see Fig. 9–9). A change in folding produces an overall change in shape which affects the surface. Why is surface so important? It is thought that molecules of various substances are able to fit onto the surface of the protein molecules with which they normally react (Fig. 9–10). In order for any reaction to occur, therefore, the fit must be a good one so that appropriate atoms of the reacting substance and the protein are brought close enough together. This depends upon the surface of the protein. It is thus possible to see that even slight changes in any one of the three levels of organization in a protein could render that protein nonfunctional. Since soluble proteins play such an important part in nearly all biochemical reactions, change in shape of these molecules can greatly disrupt metabolic processes within a cell.

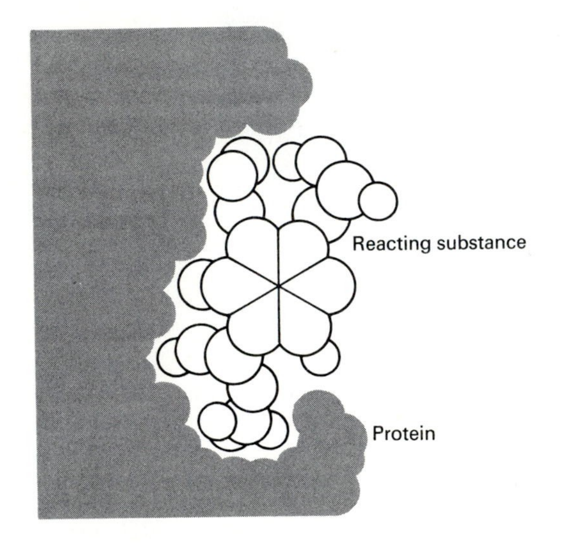

9-10
Diagrammatic representation of the close fit achieved by many reacting substances with the proteins which speed up (or catalyze) the substances' chemical changes. Surface shape of the protein is critical in allowing specific reacting molecules to fit close enough that chemical change is possible. (For more detail on the biological role of proteins as catalysts, see Chapter 10, "Enzymes.") (Margin drawing from Guttman, Biological Principles. New York: W. A. Benjamin, 1971, p. 483.)

The way in which a change in the folding of an alpha helix can affect protein specificity can be illustrated by an analogy. In Fig. 9–11, nails P, Q, R, S and T represent specific activity sites on a protein. Any change in the shape of the rubber hose, as from shape (a) to shape (b), will change the distances between one nail and the next. In this analogy, the nails represent projecting R-groups in a polypeptide chain. When the chain is folded, the R-groups are brought into various spatial patterns which give the surface of each type of protein its unique character. The spatial configuration of proteins is a striking example of the relationship between form and function on the molecular level.

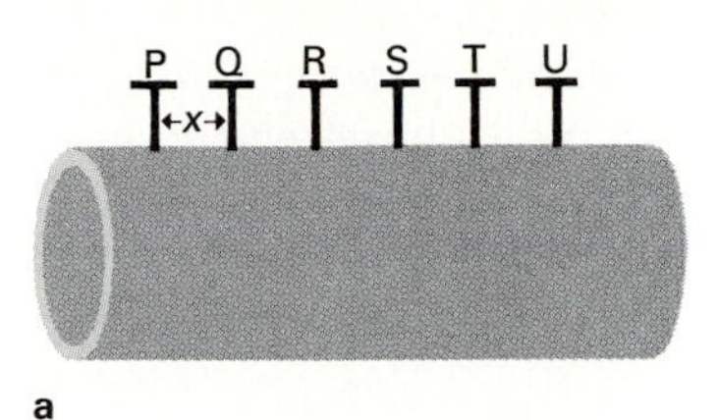

9–11
The cylinder represents a rubber tube through which nails have been driven. The nail heads in (a) above are a certain distance, *x*, apart when the tube is straight. When the tube is bent, however, the nail heads are spread further apart so that the distance between them is the greater quantity *y*. Analogously, the tubing represents the polypeptide backbone and the nails the various amino acid R-groups. If the polypeptide is bent into tertiary structure, the distances between the R-groups are changed. From (a) to (b) in the diagram represents a change in tertiary structure.

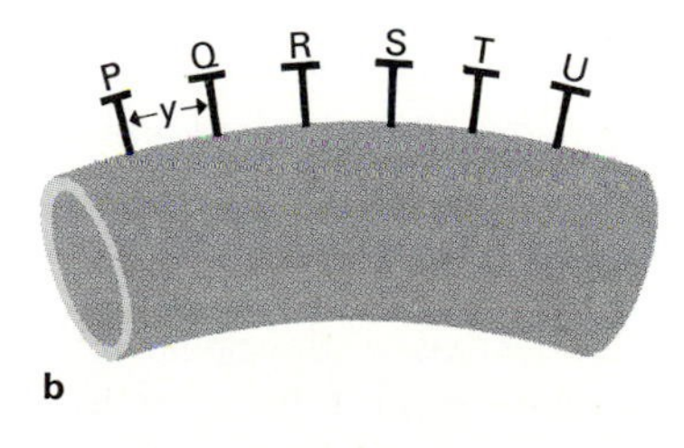

If we are to learn about the way a soluble protein functions in biochemical reactions, we must obtain the protein in an unaltered form from living cells. This is one of the most difficult problems which biochemists encounter. Proteins are very sensitive to certain environmental changes and may become irreversibly altered during the processes of extraction and purification. One of the reasons that we have been so long in learning about the three-dimensional structure of proteins has been the lack, until recently, of techniques for studying proteins without damaging their configuration. We still cannot be absolutely certain that the protein we study in a pure solution, or in crystalline form, is exactly the same as that found in living cells. Evidence indicates, however, that any differences are likely to be slight and that our present methods yield reasonably accurate information.

9–4
THE BUFFERING ACTION OF PROTEINS AND AMINO ACIDS

In solutions around pH 7, amino acids and the proteins which they compose exist in an ionized, rather than electrically neutral, form. In such solutions, the amino end of the molecule picks up an extra proton, *thus acting as a base*. Conversely, the carboxyl end loses a proton (to water, yielding a hydronium ion), *thus acting as an acid*. Under these conditions, a more correct representation of an amino acid would be:

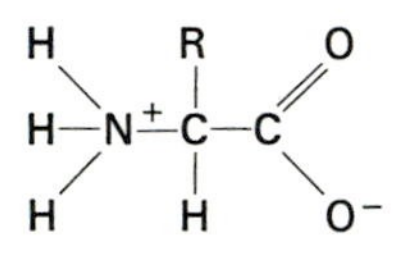

The same principle of ionization applies to whole protein molecules. No matter how long the chain of amino acids, there will always be a free amino group at one end and a free carboxyl group at the other. Thus each protein has at least one amino and one carboxyl group exposed which can ionize as shown above. In addition, many of the R-groups which project from the protein backbone, or from an individual amino acid, also contain ionizable amino or carboxyl groups (Fig. 9–12). These are also capable of acting as bases or acids by picking up or donating protons, respectively. Proteins and amino acids are thus able to act as buffers in cellular and intercellular fluids. In discussing the mechanism of this buffering in action, a single amino acid molecule will be used.

If the pH of an amino acid solution is decreased, the negatively charged carboxyl end of the molecule removes protons from hydronium ions. This is because a COO^- group is a stronger base than H_2O. Hence we have the reaction:

$$\underset{\text{(Stronger base)}}{H_3N^{+}\text{—}CHR\text{—}COO^{-}} + \underset{\text{(Stronger acid)}}{H_3O^{+}} \longrightarrow \underset{\text{(Weaker acid)}}{H_3N^{+}\text{—}CHR\text{—}COOH} + \underset{\text{(Weaker base)}}{H_2O} \qquad (9\text{–}1)$$

The carboxyl groups become electrically neutral, while the amino groups remain positively charged. In this way amino acids can serve to remove excess protons from solution.

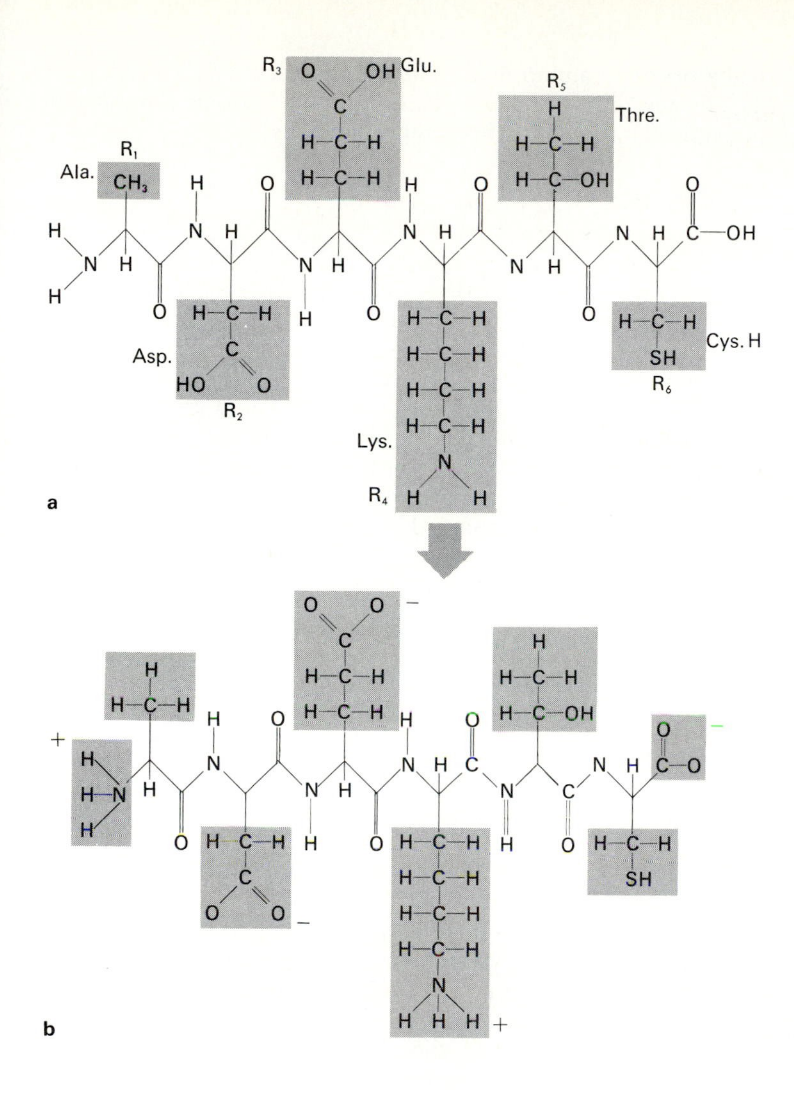

9–12

Protein chains have protruding amino (NH_2) and carboxyl (COOH) groups which, under certain conditions of pH, will gain or lose a proton. As a result, the molecule becomes polarized at the ends as well as at various points along the chain. Some R-groups do not yield or accept protons, such as R-groups 1, 5, and 6 above. These act as neither acids nor bases. By accepting or releasing protons, proteins can act as buffers in body fluids. Part (a) above shows the peptide in a non-ionized form; part (b) shows the degree to which ionization would occur at neutral pH. (See Table of Amino Acids, page 230, for abbreviations and full names of amino acids.)

If the pH of an amino acid solution is increased by the addition of OH^- or any relatively strong base, the NH_3^+ group gives up its additional proton to OH^-. This is represented as:

$$H_3N^+\text{—}\underset{\substack{|\\H}}{\overset{\substack{R\\|}}{C}}\text{—}COO^- + OH^- \longrightarrow H_2N\text{—}\underset{\substack{|\\H}}{\overset{\substack{R\\|}}{C}}\text{—}COO^- + H_2O \qquad (9\text{–}2)$$

This reaction occurs because OH^- ions are a stronger base than NH_2. Hence they tend to pull protons from the amino groups. In this way, amino acids tend to prevent the sharp rise in pH when OH^- is added to a solution. Thus we can see that amino acids and proteins can act to either add or to remove protons from solution depending upon the direction of pH change. Both reactions can be summarized in the two-way equation below:

$$\underset{\text{(Acid pH)}}{H_3N^+\text{—}\underset{\substack{|\\H}}{\overset{\substack{R\\|}}{C}}\text{—}COOH} \xleftarrow{+H^+} \underset{\text{(Neutral pH)}}{H_3N^+\text{—}\underset{\substack{|\\H}}{\overset{\substack{R\\|}}{C}}\text{—}COO^-} \xrightarrow{+OH^-} \underset{\text{(Alkaline pH)}}{H_2N\text{—}\underset{\substack{|\\H}}{\overset{\substack{R\\|}}{C}}\text{—}COO^-} \qquad (9\text{–}3)$$

Each type of protein solution has a specific pH range in which it can serve most effectively as a buffer. This difference is a result of the number of exposed amino and carboxyl groups the protein possesses. Since some amino acids have two amino groups and only one carboxyl group, or vice versa, the number of these amino acids in a protein determines its effectiveness as a buffer. The more exposed carboxyl groups a protein has, the more effectively it can pick up excess H^+ ions and thus prevent a decrease in pH. Similarly, the more exposed amino groups there are, the more effectively a protein can release H^+ ions into them and thus prevent an increase in pH.

It can now be understood how a drastic change in pH can alter the shape of (*denature*) a protein. When the exposed ionizable groups are subjected to a different pH, they either gain or lose protons (depending upon the direction of the pH change), and this in turn causes the pattern of hydrogen bonds to change. Since hydrogen bonds determine to a large extent the secondary structure of a protein (refer to the footnote on page 162), a change in hydrogen bond formation may well lead to a change in protein configuration. (This can be demonstrated by dropping egg albumen into a concentrated HCl solution.)

9–5
HOW IS PROTEIN STRUCTURE DETERMINED?

Working out the complete three-dimensional structure of proteins is a feat which has been accomplished only in the past 15 years. The task involves first determining the amino acid sequence. Then, the portions of the polypeptide chains which exist as alpha helix must be charted. Finally, the folding of the polypeptide chain, or chains, must be determined.

The complete amino acid sequence is very difficult to determine. It was first shown for one of the smallest proteins, the hormone *insulin*. Figure 9–13 shows the complete amino acid sequence for the two polypeptide chains of insulin. The sequence is determined by

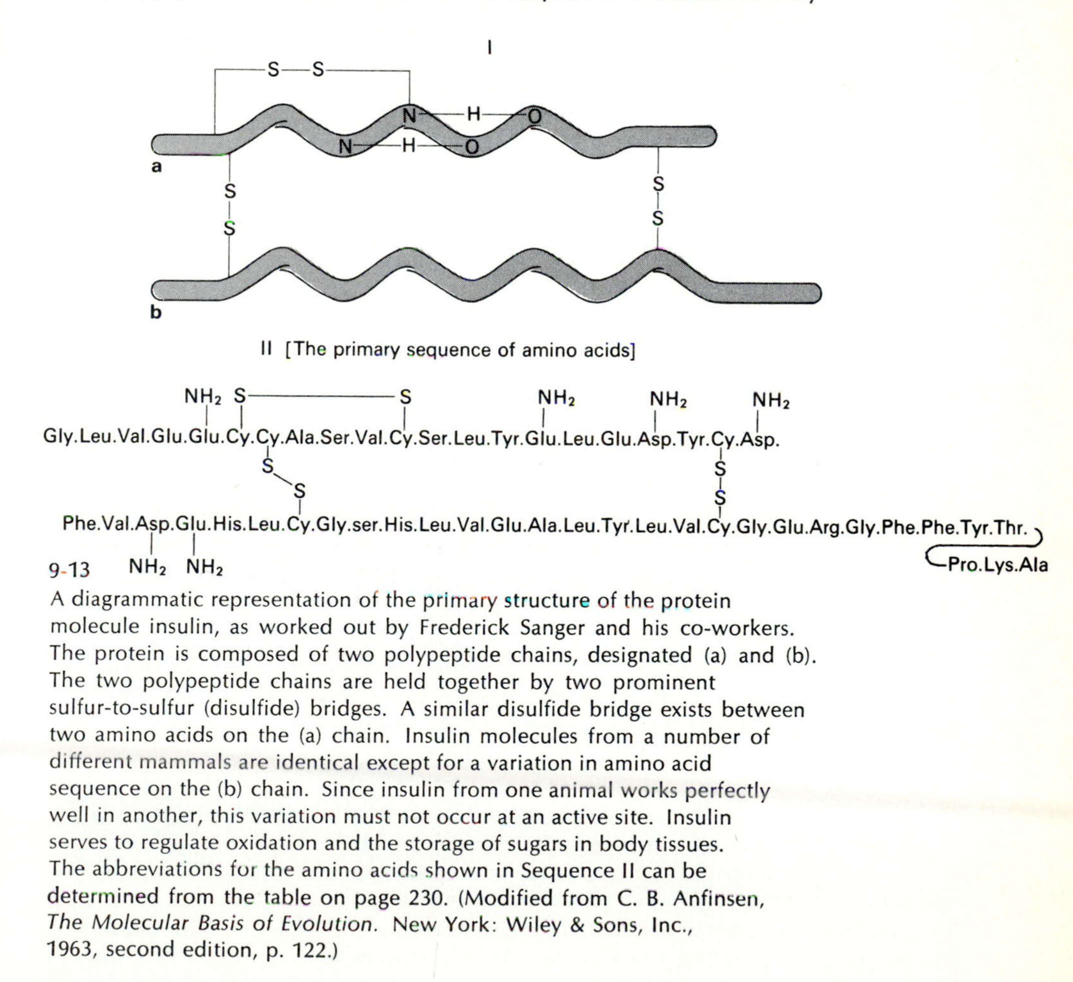

9-13
A diagrammatic representation of the primary structure of the protein molecule insulin, as worked out by Frederick Sanger and his co-workers. The protein is composed of two polypeptide chains, designated (a) and (b). The two polypeptide chains are held together by two prominent sulfur-to-sulfur (disulfide) bridges. A similar disulfide bridge exists between two amino acids on the (a) chain. Insulin molecules from a number of different mammals are identical except for a variation in amino acid sequence on the (b) chain. Since insulin from one animal works perfectly well in another, this variation must not occur at an active site. Insulin serves to regulate oxidation and the storage of sugars in body tissues. The abbreviations for the amino acids shown in Sequence II can be determined from the table on page 230. (Modified from C. B. Anfinsen, *The Molecular Basis of Evolution*. New York: Wiley & Sons, Inc., 1963, second edition, p. 122.)

breaking down the two polypeptide chains so that fragments of different sizes are obtained. This is done by using enzymes known to break polypeptide chains only at specific points. By analyzing the content and sequence of amino acids in these fragments, it is possible to find where each fragment fits in relation to all the others. In this way, the primary structure of the entire protein molecule is finally determined.

Once the primary structure is determined, the secondary and tertiary structure can be determined by physical means. X-ray diffraction (Section 7–3) is one such technique. The work of J. C. Kendrew and others has shown the complete geometrical pattern for myoglobin and ribonuclease.

9–6 SUMMARY

Proteins serve living organisms both as enzymes and as structural elements. The ability of proteins to carry out very specific reactions is the result of their primary, secondary, and tertiary structure.

It is difficult to overestimate the importance of proteins to living systems. An organism's carbohydrates and lipids are much like those of any other. Its proteins, on the other hand, are unique. Why they are unique, and how they got that way, is the topic of the next two chapters.

PROTEIN DENATURATION—WHO NEEDS IT?

It has been stated several times that the shape of a molecule may determine its function. This is especially true for proteins—their shape is essential to biological function. Sometimes, however, the loss of shape in denaturation of proteins may serve a useful purpose. Following are a few examples.

When food enters the stomach, hydrochloric acid is secreted by the stomach wall. Hydrochloric acid is a very strong acid—approximately pH 2 in the stomach—and it will cause denaturation of many proteins that come into contact with it. This is especially evident when a baby burps and a little bit of the curdled material—denatured milk proteins—is coughed up with the burp. The advantage of curdling milk proteins is to cause precipitation in the stomach rather than to let the fluid pass through undigested. The semisolid precipitate formed by the denatured proteins must now pass through the digestive tract more slowly, allowing more complete digestion by making the food more accessible to degradation by the enzymes of the intestine; thus the nutritional process is improved.

Physical force can also be used to denature proteins. For example, when liquid heavy cream is beaten, it will become thick and firm enough to make peaks (whipped cream); further beating will cause the formation of butter.

Chapter 10
Enzymes

10–1 THE NATURE OF CATALYSTS

There are many varieties of proteins in the bodies of living plants and animals. However, one group deserves special attention: the *enzymes.* Enzymes are involved in virtually all of the chemical reactions within living organisms. Without these specialized proteins, life as we know it could not possibly exist.

If potassium chlorate ($KClO_3$) is put into a test tube and heated to 360°C, oxygen is given off. Eventually, only potassium chloride remains.

However, if a small amount of manganese dioxide (MnO_2) is added to the reaction, oxygen is liberated at only 200°C. Furthermore, the reaction proceeds at a faster rate. Chemical analysis of the end products shows that the potassium chlorate has again become potassium chloride. However, *the manganese dioxide is still chemically unchanged!*

What has actually happened in this reaction can only be hypothesized. However, this is not a unique case. There are many chemical reactions which are affected by the addition of another element or compound. The rate of the reaction is changed. The properties of the added element or compound causing this effect seem unchanged after the reaction has occurred. In such cases, the added element or compound is called a *catalytic agent,* or a *catalyst.*

Enzymes are organic catalytic agents. They are present in almost al! of the chemical reactions which keep organisms alive. They enable a human being to be a beehive of chemical activity at only 98.6°F. They enable Antarctic fish to remain alive and active at close to 0°C. Furthermore, since they come out of chemical reactions unchanged, they can be used over and over again. This means that an organism does not have to expend a great deal of energy in order to constantly resynthesize enzymes at a rate proportional to the rate of the reactions they catalyze.

The compound upon which an enzyme acts is called its *substrate.* Thus the substrate for sucrase is sucrose; for lactase, lactose; for trypsin, proteins; and so on. There is often a wide difference in molecular size between enzymes and the substrate on which they act. For example, the enzyme sucrase, being a protein, has a molecular weight of many thousands. Sucrose, on the other hand, contains only 45 atoms. Its molecular weight is only 342. Enzymes are thus considerably larger in size than the substrate molecules on which they act.

10–2 OBSERVATIONS ON ENZYME-CATALYZED REACTIONS

Much can be learned about the nature of enzyme-catalyzed reactions by studying the kinetics of reactions: i.e., the various changes in rate which a reaction shows during its course. For example, it is possible to measure the amount of product formed in an enzyme-catalyzed reaction from the moment the reactants (called the "substrate") and enzyme are brought together, until the reaction has stopped (i.e., reached an end-point, or equilibrium).

If the amount of product formed is measured at one-minute intervals and this quantity is plotted against time on a graph, curves such as those shown in Fig. 10–1 are obtained. This graph shows data for two different reactions, designated here as A and B. Observe the solid line for reaction A as an example. At time 0 there is no product detectable in the reaction system; this represents the beginning of the reaction. After 1 minute 20 μmoles have been formed, after 2 minutes 40, after 3 minutes 60, and after 4 minutes 80. The rate of the reaction could be given as 20 μmoles of product formed per minute for this initial period. Note, however, that by the fifth minute this

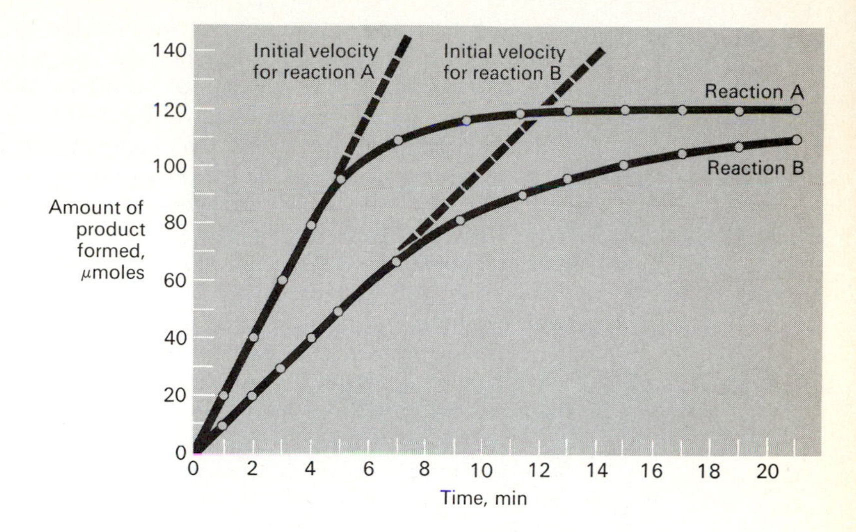

10-1
The relationship between total amount of product formed and the time for two enzyme-catalyzed reactions. The initial rates of reactions A and B are different, measuring a difference in turnover number. As shown in this figure, A acts more effectively on its substrate (in a given period of time) than does B (it is assumed that both reactions will ultimately reach the same equilibrium, i.e., same concentration of reactants and products). If both reactions begin with the same concentration of substrate, eventually equivalent amounts of product will be formed (and the curves will level off at the same height). Only the time required to yield the same total amount of product will differ.

rate has begun to slow down. Instead of a total of 100 μmoles of product formed by the five-minute mark, only about 92 have been formed. In the interval between the fourth and the fifth minute only about 12 μmoles, rather than 20, have been formed. During the first four minutes, the rate has been constant, and the change in rate has been zero. But from the fourth minute on through about the twelfth minute, the rate is changing; i.e., it is slowing down. For each successive minute, the amount of product formed in that interval is less than in the preceding minute. From the twelfth minute onward, the reaction rate again becomes constant; no more product is being formed, and the change in rate is again zero.

Using the principles of chemical kinetics discussed in Chapters 4 & 5, how can we explain these changes? During the first four minutes of reaction A, the number of substrate molecules greatly exceeds the number of enzyme molecules (this is usually true for all catalyzed reactions). This means that every enzyme molecule is working at its maximum capacity. There are so many substrate molecules around that, each time an enzyme finishes with one substrate molecule, it is confronted with others. After a period of time, however (in this case, after four minutes), the number of substrate molecules begins to dwindle and the concentration is lowered. As the concentration of reactants (substrate) is lowered, the chance of successful collisions is reduced (see Section 4–5). In the present case we are concerned not only with successful collisions between the reactants, but also between reactants (substrate) and the enzyme. As time goes on, the substrate concentration becomes less and less since more substrate is being converted into product. The leveling off of the curve after twelve minutes indicates that the total amount of product is remaining constant. The reaction system has reached equilibrium.

A similar curve can be seen for reaction B. The difference in the shape of this curve represents the difference in effectiveness of enzyme A and enzyme B in acting on their respective substrates. Given the same initial concentrations of both enzyme and substrate, it is apparent that enzyme B acts less rapidly on its substrate than enzyme A. Since the initial substrate concentrations in the two reaction systems are the same, the total amount of product formed in B, given enough time, will eventually be equivalent to that formed in A, as long as the two reactions have the same equilibria—i.e., as long as the reactions normally reach equilibrium with the concentration of reactants and products the same for the two reactions.

In comparing the kinetics of reactions A and B, it is necessary to have a common reference point. For example, do we want to compare the two reactions at the two-minute mark, the twelve-minute mark, or later? It is obvious that comparisons made at different times will give different values for the total product formed in reactions A and B. The reason for this is that we are faced with two variables (given equal concentrations of enzyme molecules): (1) the rate at which the two different enzymes can act on their substrates (which is a characteristic of the enzyme molecule); and (2) the constantly changing substrate concentrations during the course of reaction. There is, however, a means of eliminating one of these variables. In the first few minutes of the reaction, the number of substrate molecules is so large compared to the number of enzyme molecules that changing the concentration does not, for a period at least, affect the number of successful collisions. Note that during this early period the rate of change is constant; i.e., the enzyme is acting on substrate molecules at a constant rate. The slope of the graph line during these early minutes defines what biochemists call the *initial velocity* of the reaction. The initial velocity of any enzyme-catalyzed system is determined by the characteristics of the enzyme molecule and is always the same for the enzyme and its substrate as long as temperature and pH are constant and substrate is present in excess. As shown in Fig. 10–1 the initial velocities of reactions A and B are different and represent a difference in effectiveness between the two enzymes. This difference is measured as the *turnover number,* a term which indicates the maximum rate at which an enzyme can act on substrate molecules in a given period of time. It is important to understand why substrate should be present in excess to measure the true initial velocity of an enzyme-substrate reaction.

Suppose that we run a series of reactions with enzyme system A, in which the concentration of enzyme is held constant but the starting concentration of substrate is varied. If we measure the initial velocity (say, for the first two minutes) of each reaction and plot this value on a graph against substrate concentration, we get a curve like that shown in Fig. 10–2. Note that at low substrate concentrations, the initial velocity of the reaction is low. This means that when there are few substrate molecules in solution to begin with, the enzyme can never reach its maximum rate of conversion (since the frequency of successful collisions becomes the *limiting factor*). In other reactions, where the starting concentration of substrate is greater, the initial velocity is greater. As the graph line shows, the increase in initial velocity with increasing substrate concentration is linear, but only up to a point. By the time the starting concentration of substrate has reached 0.1 mole per liter, the curve has begun to level off. This suggests that the enzyme is approaching its maximum initial velocity. In other words, beyond about 0.5 moles per liter an increase in substrate concentration does not affect the initial velocity of the reaction. Enzyme molecules are working as fast as they can; the presence of more and more substrate at the start of the reaction will not affect the initial velocity. Thus, at this point we can measure true initial (maximum) velocity because one variable has been eliminated: substrate concentration. Rate of conversion now depends solely on characteristics of the enzyme molecule.

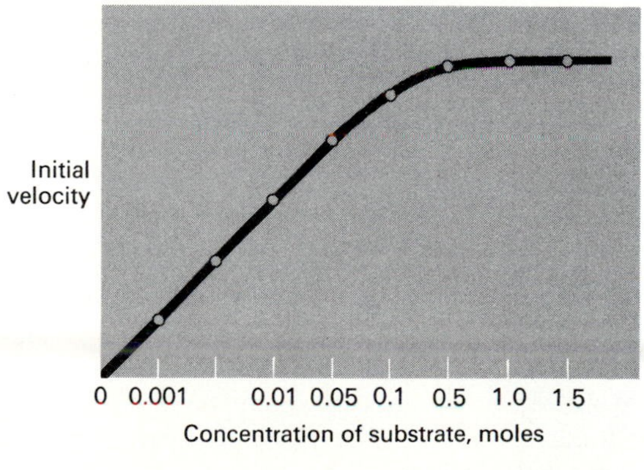

10-2
Graph showing the relationship between substrate concentration and initial velocity of an enzyme-catalyzed reaction. The reaction volume and enzyme concentration are held constant. Each point on the graph represents the measured initial velocity of a specific reaction where a substrate concentration was the only variable.

The above phenomena are understandable in terms of our general knowledge of the kinetics of any chemical reactions, whether enzyme-catalyzed or not. However, enzyme-catalyzed reactions have some characteristics which are distinct from non-enzyme-catalyzed reactions. Recognition of the nature of some of these characteristics will help to understand a second important question: By what mechanism do enzymes speed up the rate of biochemical reactions?

1) *The rate of enzyme-catalyzed reactions is greatly affected by temperature.* If the initial velocity of a specific enzyme-catalyzed reaction is measured at a number of different temperatures (with enzyme and substrate concentrations held constant), a curve like that shown in Fig. 10–3 is obtained. Note that at low temperatures the rate of reaction is quite slow, and that with increasing temperatures, up to about 37°C, the rate increases. Beyond 37°C, however, the rate begins to slow down again even though the temperature is raised. From our knowledge of all chemical reactions, we can explain the first half of this curve: why the initial velocity increases with an increase in temperature. The higher the temperature, the more rapidly the reacting molecules (in this case, substrate and enzyme molecules) move about, and the greater the fraction of molecules which possess minimum energy of activation. However, this principle should apply to temperatures above 37°C as well. In non-enzyme-catalyzed reactions, initial velocity does indeed increase as the temperature is raised beyond 37°C. Why, then, are enzyme-catalyzed reactions so sensitive? To answer this question requires that we develop a model to explain how enzymes function, a topic that will be taken up in the next section.

The initial velocity-temperature curve shown in Fig. 10–3 varies from one type of enzyme to another. Every specific enzyme has its so-called *optimum temperature:* that temperature at which the enzyme achieves its maximum rate (i.e., initial velocity). Thus, for the particular reaction shown above, 37°C is the optimum temperature; for another enzyme-catalyzed reaction, 25° or 40° might be the optimum.

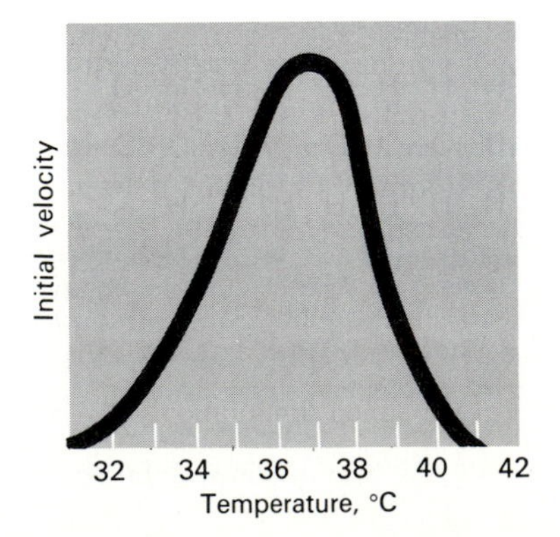

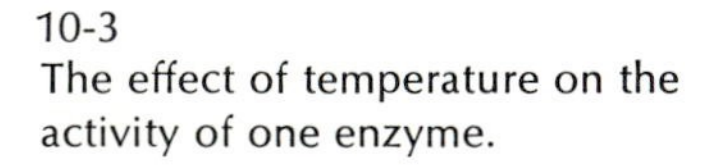
10-3
The effect of temperature on the activity of one enzyme.

The term optimum temperature is somewhat misleading and should be viewed not only in a chemical but also a biological context. Seldom in nature do biochemical reactions operate at their maximum rate. It is not often advisable, for example, for an organism to be carrying out a particular biochemical reaction as rapidly as possible. Thus, while a particular enzyme may work at a maximum rate at 38°C, we cannot assume that this is the temperature at which the enzyme usually functions. The enzyme could normally exist inside a mammal which maintains a fairly constant internal temperature of 37°C, or it could exist in a bacterium which lives in the soil, where the temperature varies considerably. In general, the optimum temperature of an enzyme-catalyzed reaction, as measured in a chemical sense, is often close to the average temperature at which the enzyme functions in nature. But we must not assume that this is always true. And, more important, we must not assume that whatever temperature (or other conditions) may be optimal when the system is measured in a test tube is the "optimum" at which the system operates in terms of the survival of organisms. The fastest rate is not always the best from the standpoint of an organism's operating efficiency.

2) *Enzymes can be "poisoned"* by certain compounds, such as bichloride of mercury or hydrogen cyanide. These chemicals are deadly poisons to all living organisms. They exert their effect by inactivating one or many enzymes. Hydrogen cyanide blocks one of the enzymes involved in the chemistry of respiration. The way in which this is believed to occur will be considered shortly.

3) *Enzymes are specific in their action.* This is one of the most distinctive characteristics of enzymes. They will often catalyze only one particular reaction. For example, the enzyme sucrase will catalyze only the breakdown of sucrose to glucose and fructose. It will not split lactose or maltose. Lactase and maltase, respectively, must be used as the catalytic agent for these two sugars.* With other enzymes, specificity of action is not quite so obvious. Trypsin, for example, is a *proteolytic* (proteïn-splitting) enzyme which acts upon many different proteins.

Why should some enzymes work on only one compound while others will work on several? Enzymes act upon a specific chemical linkage group. In the case of trypsin, only those peptide linkages of

---

* The *-ase* ending indicates that the compound is an enzyme. Other enzymes, such as trypsin, end with *-in*. This signifies that they, like all enzymes, are proteins. Enzymes ending in *-in* were discovered and named before an international ruling was made in favor of the *-ase* ending. A few changes have been made. For example, the mouth enzyme ptyalin is now called salivary amylase. Enzymes are also named after the compounds they attack. Thus peptides are attacked by peptidases; peroxides by peroxidases; lipids by lipases; ester linkages by esterases; hydrogen atoms are removed by dehydrogenases; and so on.

protein molecules which are formed with the carboxyl group of the amino acids lysine or arginine are acted on by trypsin:

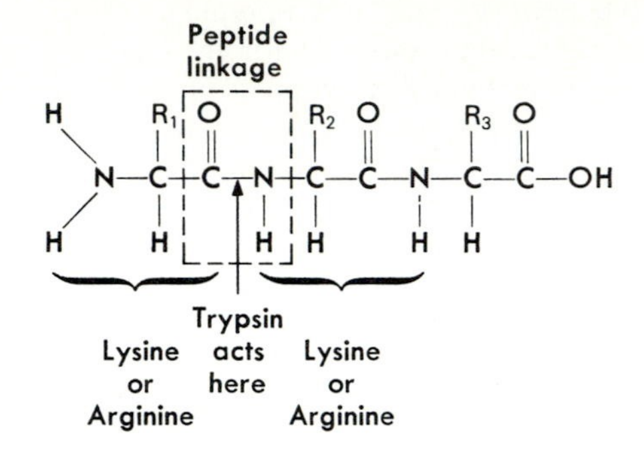

Since peptide linkages involving the carboxyl groups of lysine and arginine are characteristic of many proteins, it is not surprising that trypsin will act on more than one.

10–3 THEORIES OF ENZYME ACTIVITY

We now come to one of the most important questions in the study of enzymes: By what mechanism do enzymes operate?

The most important hypothetical model proposes that enzymes have certain surface configurations produced by the three-dimensional folding of their polypeptide chains. On this surface, there is an area to which the substrate molecule is fitted. This area is called the **active site** of the enzyme. It is thought that, when the substrate molecule becomes attached to the enzyme at this site, the internal energy state of the substrate molecule is changed, bringing about the reaction (Fig. 10–4).

A helpful analogy for visualizing how enzymes work is to picture the substrate as a padlock and the enzyme as the key which unlocks it. The notched portion of the key thus becomes the active site, since it is here that the "reaction," or the unlocking of the padlock, takes place. The padlock comes completely apart, just as a molecule is broken apart by enzyme action. The key serves equally well, however, to run the reaction in the reverse direction, i.e., to lock the padlock again. The key comes out unchanged and ready to work again on another padlock of the same type; similarly, the enzyme is ready to catalyze another reaction of the same type. In light of this analogy, trypsin becomes a sort of "skeleton key" enzyme. It can open several types of padlocks (proteins) as long as they have similar engineering designs (certain peptide linkages).

The lock-and-key model has been nicely supported by X-ray diffraction studies. The structure of the complex that results when the

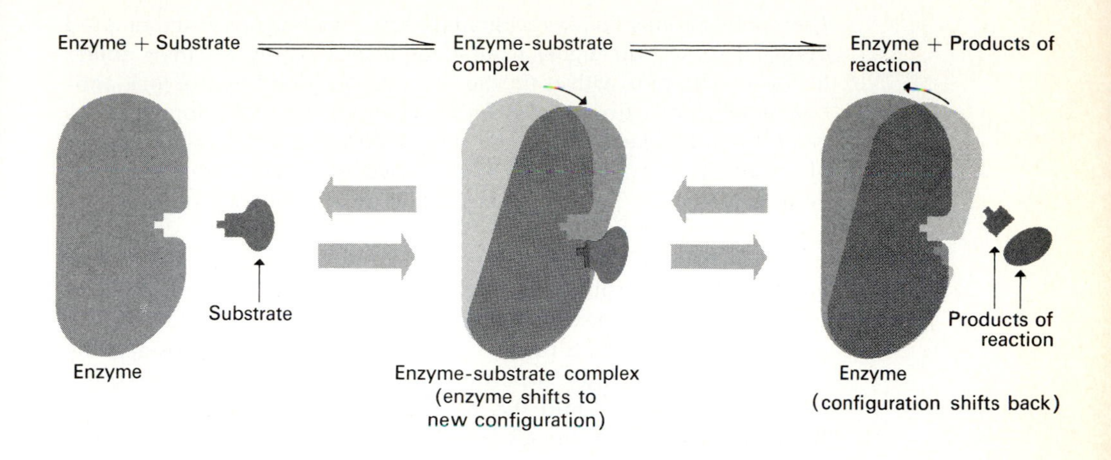

10-4
Schematic version of the interaction of enzyme and substrate. The specificity of enzymes for certain substrates is thought to be due to the surface geometry of the enzyme and substrate molecules, which allows them to fit together in a precise manner. Once the substrate molecule is situated in the appropriate surface region of the enzyme, certain groups of atoms in each molecule interact in such a way that the substrate is permanently changed. Enzymes can speed up the rate of breakdown (or synthesis) of substrate molecules. The enzyme molecule is only temporarily altered by interaction with the substrate, and can thus catalyze other substrate molecules as soon as it finishes with one.

enzyme lysozyme reacts with a molecule very similar to its normal substitute shows that the substitute "substrate" molecule fits snugly into a groove or cleft in the lysozyme molecule's surface, where the enzyme's catalytic groups are held in just the right position to break the substrate's bonds. Yet there is some evidence that the lock-and-key model of enzyme action is not completely satisfactory in terms of always yielding accurate predictions with all enzymes. For example, the enzyme specific for the amino acid isoleucine must often "choose" between its proper substrate (isoleucine) and valine. The only difference between these two molecules is one methylene (—CH_2—) group:

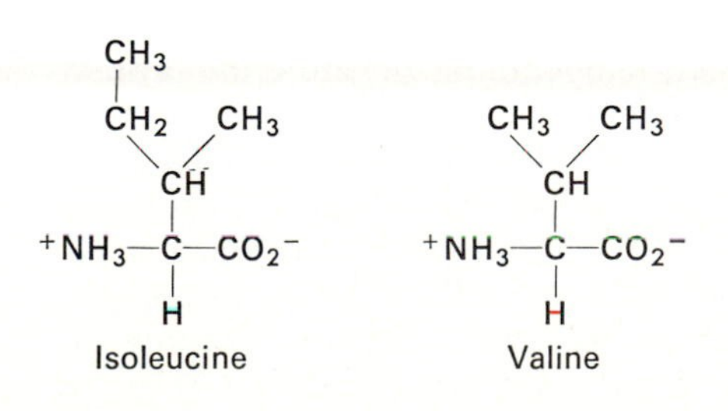

The chemist Linus Pauling estimates that if an enzyme discriminated between these two highly similar molecules purely on their ability to form a complex with it (in the manner described by the lock-and-key model), then from these physicochemical grounds alone the enzyme should make a "mistake"—i.e., it should bind to valine instead of isoleucine—about 1 time in 20. However, it has been shown experimentally that "mistakes" are made at a frequency of less than 1 in 3000.

This difficulty can be overcome if many enzymes are looked upon as being "flexible" rather than rigid. When proper substrates bind to these enzymes, a change is induced in the structure of the latter which results in a reorientation of the enzyme groups actually involved in the catalysis. In other words, the enzymes' "active sites" portions are brought into the proper position for action. The "induced fit" hypothesis suggests that enzyme specificity is only partly due to complementarity of structure between enzyme and substrate molecules, and that the ability of substrates to induce in the enzyme the structural changes necessary for catalysis must also be taken into account.

Recent X-ray diffraction work on the proteolytic (protein-splitting) enzyme carboxypeptidase A provides very strong evidence in support of the "induced fit" hypothesis. It has been shown that the binding of this enzyme to its substrate causes movement of the side-chain amino acid number 248, tyrosine (known to be involved in the actual catalysis), some eight angstroms toward the substrate, so that tyrosine's hydroxyl (—OH) group is near the substrate peptide bond to be split (see Fig. 10–5). Furthermore, the positively charged group of the arginine (145) moves two angstroms toward the substrate carboxyl group, where it binds it. It has further been shown that an inhibitor of carboxypeptidase A binds to the enzyme as does the normal substrate, but does not bring about a change in the tyrosine (248) side-chain.

The lock-and-key analogy, including the "induced fit" hypothesis, helps us to interpret the characteristics of enzyme action described in the previous section; for example, refer to Fig. 10–3. The effect of increase in initial velocity with increasing temperature (up to 37°C) is a result of increasing the kinetic energy of the substrate and enzyme molecules. At temperatures higher than 37°C, however, the decrease in initial velocity is the result of changes in the configuration of the enzyme molecule, a protein. We know that most proteins are heat sensitive. High temperatures denature the proteins by breaking hydrogen bonds or other types of bonds which hold the molecule in its specific three-dimensional shape. When the enzyme's shape is altered, the active site no longer fits the specific configuration of the substrate molecule. The result is that no reaction can occur. Thus, when an enzyme system is exposed to increasing temperatures, more and more of the enzyme molecules become denatured. Though with increasing temperature more molecules are

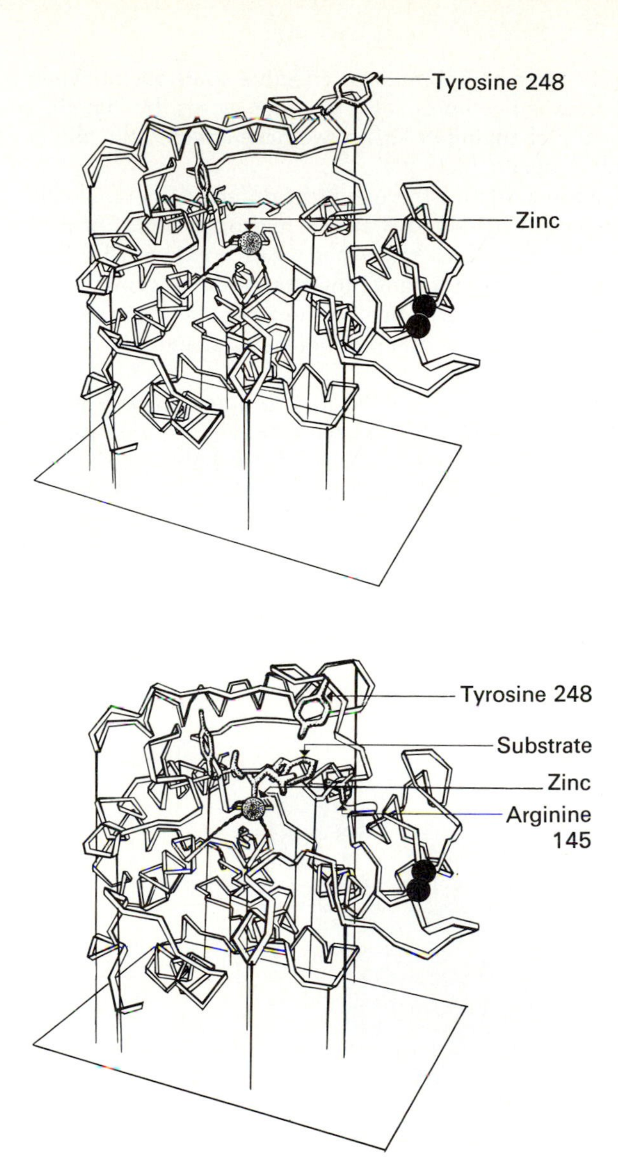

10–5
At top is shown a drawing of a three-dimensional model of the enzyme carboxypeptidase A, as reconstructed from x-ray diffraction data. The groove or cleft into which the substrate molecule fits is shown above and to the right of the zinc atom. At bottom is shown a drawing of the same enzyme, but with the substrate molecule bound to it at the zinc atom and a side-group of arginine. After the substrate molecule is in position, both tyrosine and arginine move toward it. This observation supports the "induced fit" hypothesis of enzyme action illustrated in abstract model form in Fig. 10–4.

colliding with each other, the number of effective collisions between enzyme and substrate is becoming less. In other words, by "tampering" with the key and changing its shape, we have affected the ability of the key to open a specific lock.

A similar situation exists for the effects of different pH ranges on the action of enzymes. Changing pH affects hydrogen bonds of proteins; at high or low pH values many hydrogen bonds are broken, and thus the enzyme molecules change their three-dimensional structure. The effect of both pH and heat on enzymes emphasizes the importance which preserving a specific molecular shape has on the biochemical activity of enzymes.

Especially interesting in light of the lock-and-key analogy is the effect of inhibitors, or "poisons," on enzyme activity. A very simple case, diagrammed in Fig. 10–6 will illustrate this problem. In almost all cells glucose may be stored indefinitely without ever releasing its chemical energy. The reason for this is that there is no activation energy present. In terms of the energy hill, there is nothing to lift the glucose over the hump at the top and start it rolling down the hill. Outside of a living system, it is necessary to supply a relatively large amount of activation energy (heat, for example) to glucose in order to get it started down the energy hill.

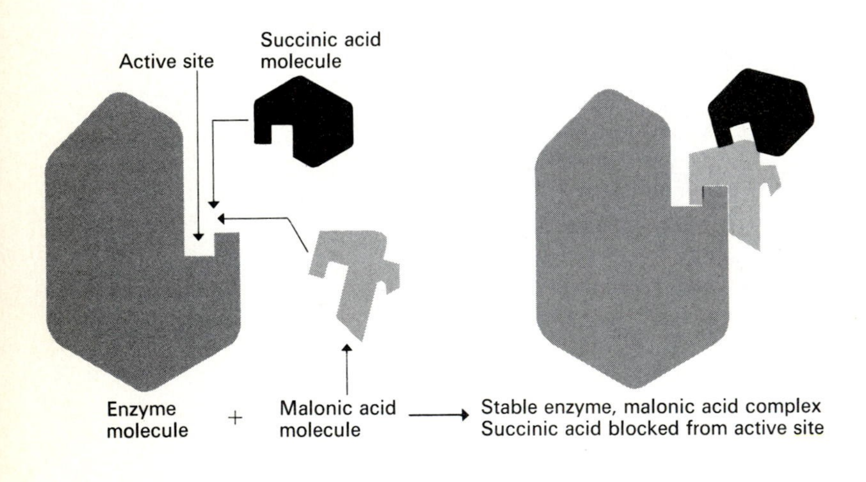

10–6
A diagrammatic representation of competitive inhibition in enzymatic reactions.

One step in the breakdown of sugars involves the conversion of a four-carbon molecule, succinic acid, to another four-carbon molecule, fumaric acid, with the removal of two electrons (and two protons). This reaction is catalyzed by the enzyme succinic dehydrogenase, which is highly specific for its succinic acid substrate. If the reaction is run in a test tube with just enzyme and substrate, a particular initial velocity can be observed; if malonic acid is added to the test tube along with succinic acid, the rate of formation of fumarate is greatly reduced. Malonic acid is a three-carbon molecule whose overall molecular shape is very similar to that of succinic. It is thought that the malonic acid molecule can fit into the active site of succinic dehydrogenase and "fool" the enzyme. However, because malonic acid is slightly different from succinic, the enzyme cannot convert it into fumaric. Hence, whenever a malonic acid molecule gets into the active site of succinic dehydrogenase, no reaction occurs. Malonic acid inhibits the enzyme system, a feat which it accomplishes by virtue of its similar shape to the normal substrate. Malonic acid is like a key which fits into a lock but which is just different enough from the proper key not to be able to turn. For a moment it "jams" the lock. Malonic acid molecules can fall out of the active site, however, and, if they are around, it is possible for succinic acid molecules to enter once the site is free. Malonic acid is thus an example of a *competitive inhibitor:* it competes with the normal substrate for the enzyme's active site. However, it does not permanently inactivate the enzyme.

Certain molecules can also affect enzyme molecules more or less permanently. These are called *noncompetitive inhibitors,* two examples of which are carbon monoxide and cyanide. Carbon monoxide molecules attach to the active site of certain oxygen carriers (e.g., hemoglobin) and certain respiratory enzymes (e.g., the cytochromes). When either of these inhibitors is attached to a respiratory protein, normal function is impossible (i.e., the cytochrome or hemoglobin cannot interact with oxygen). Unlike malonic acid, however, neither carbon monoxide nor cyanide becomes detached from the active site of the protein. As a result, the effect of noncompetitive inhibitors is permanent; this is why both carbon monoxide and cyanide are such deadly poisons.

Since most biochemical reactions are to one degree or another reversible, it is not surprising that enzymes can catalyze reactions in either direction. In a completely reversible reaction, the enzyme can catalyze the reverse reaction as readily as the forward. If the equilibrium is shifted to the right, the enzyme can catalyze the forward direction more easily than the backward, and so forth. This observation suggests the aforementioned point that enzymes do not make reactions occur which would not occur on their own, but only increase the *rate* at which the reactions take place. *Enzymes speed up reactions by changing the energy requirements for getting the reaction started.* Enzymes do not affect the net energy changes (the ΔF_0 of any reaction; that would be thermodynamically impossible.

How, then, do enzymes increase reaction rate? The energy hill analogy, discussed earlier in Chapter 4, will help to elucidate this point. The glucose molecule at the top of the energy hill shown in Fig. 10–7 represents a certain amount of potential chemical energy.

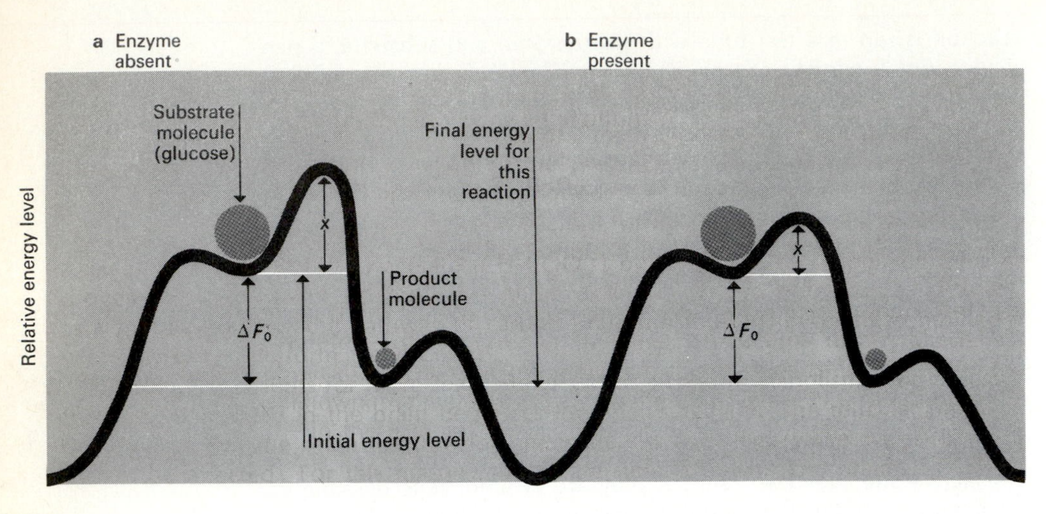

10-7
The effect of an enzyme on the activation energy requirements of a molecule undergoing chemical degradation. The amount of activation energy required is represented by the distance x. The net gain in free energy is symbolized by F_0 (it is assumed that the molecule falls to the lowest energy level). Note that the uncatalyzed reaction (a) has a far higher activation energy barrier than the catalyzed reaction (b) on the right. The new molecule must overcome another energy barrier, requiring more activation energy, before it can be broken down and release more energy. In a living organism another enzyme, specific for this reaction, would be needed.

However, glucose may be stored indefinitely without ever releasing its chemical energy. In terms of the energy hill, there is nothing to lift it over the hump at the top and start it rolling down the hill.

This is where enzymes fit into the picture. Their presence lowers the amount of activation energy (x) needed to start the reaction. Thus by expending only a small amount of energy, a living organism can release the chemical energy available in the glucose molecule.

Many enzymes, besides requiring environmental conditions such as proper temperature, pH, etc., also need the presence of certain other substances before they will work. For example, salivary amylase will work on amylose only if chloride ions (Cl^-) are present. Magnesium ions (Mg^{++}) are needed for many of the enzymes involved in the breakdown of glucose.

Some enzymes require another organic substance in the medium in order to function properly. In a few cases, enzymes actually consist of two molecular parts. One of these is a protein, called an *apoenzyme*. The other molecular part is a smaller, nonprotein molecule. This smaller molecule is called a *coenzyme*. Its name signifies that it works with the main apoenzyme molecule as a coworker in bringing about a reaction.

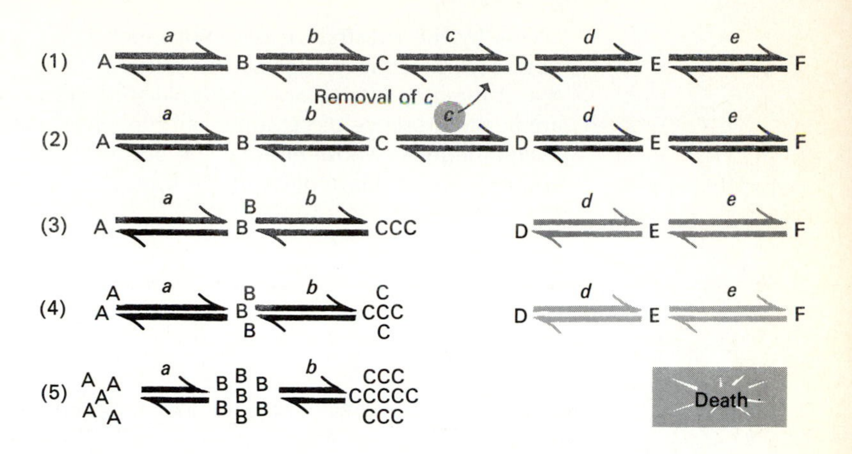

10-8
(1) A series of six chemical reactions, each catalyzed by its own specific enzyme (*a, b, c, d,* and *e*.) (2) Enzyme *c* is removed, preventing the conversion of C to D. (3) Since the reactions are reversible, A, B, and C begin to accumulate. Since no more D, E, or F is being produced, they begin to disappear as F is used by the organism in its life processes. (4) Further accumulation of A, B, and C and depletion of D, E, and F. (5) Death of the organism due to lack of the vital substance F. Had the reactions been irreversible, there would have been accumulation of only C, continuing until the death of the organism. This demonstrates the far-reaching effects of the removal of one important enzyme and the dynamic chemical balance which exists in living systems.

In an apoenzyme-coenzyme case, the two molecular parts are chemically bonded to each other. In other cases, the coenzyme is combined only briefly with the enzyme. In either case, the presence of the coenzyme is needed before any catalytic activity takes place.

Chemical analysis of the smaller coenzymes has shown that they often contain a vitamin as part of the molecule. This finding has led to the idea that vitamins serve as coenzymes. This would explain why the absence of certain vitamins causes such remarkable physical effects on the organism. The enzyme which works with the vitamin-based coenzyme cannot work by itself. Therefore, an entire series of important physiological reactions may be blocked (see Fig. 10–8). The hypothesis that vitamins serve as coenzymes also explains why only a small supply of vitamins is sufficient to fulfill the requirements for good health. Like enzymes, coenzyme molecules must be replaced only from time to time, at a relatively slow rate.

Small molecules such as coenzymes could affect enzyme activity by interacting with specific active sites (not those for the substrate necessarily) on the enzyme molecule. It is thought that such an interaction could cause the three-dimensional shape of the enzyme to shift slightly, thus opening up the substrate active sites for easier access by the substrate itself. Not much is known at present about how accurate this model is, but certain kinds of evidence suggest that something like a configurational shift does occur when enzymes are in the presence of certain small molecules such as coenzymes, or even inhibitors (whose effect is the opposite from coenzymes). More will be said on this subject in the following chapter.

COMMERCIAL USE OF ENZYMES

Enzymes have been used in recent years for a number of commercial purposes. One of the most familiar is their use as meat "tenderizers." The tenderizer known as "Adolf's" contains papain, a generalized proteolytic (protein-splitting) enzyme extracted from certain plants. The function of the enzyme in tenderizers is to begin digesting connective tissue (the white, sinewy portion of meats), which contributes significantly to toughness. Papain has very broad specificity—that is, it catalyzes the breakdown of many proteins. Hence, as a tenderizer it also digests away part of the muscle tissue as well as the connective tissue. For this reason only small amounts of tenderizer are added to meat, and once the tenderizer is added the meat must be used reasonably soon. The tenderizer literally "predigests" the meat before it ever gets into a consumer's stomach!

Enzymes are also used for cleaning purposes. The protein-digesting enzyme subtilisin (extracted from the bacterium *Bacillus subtilis)* has been incorporated into several presoak laundry agents and some laundry detergents. Subtilisin has an active site very similar to several other protein-digesting enzymes (such as trypsin) found in animal digestive tracts, and it is particularly effective in removing protein-containing stains (such as chocolate or coffee) in clothes. Subtilisin has the added advantage of being very heat-stable, so that it can continue to act even under high temperatures in a washing machine. Like papain, however, subtilisin will digest a broad range of proteins, including not only stains but silk or cotton *fabric* as well. The use of subtilisin or other enzyme-containing laundry agents thus shortens the life of most fabrics considerably.

ALKAPTONURIA: AN INBORN ERROR OF METABOLISM

The possibility of affecting the dynamic chemical equilibrium by the absence or lack of function of an enzyme in a series of chemical reactions was discussed briefly on page 191. An example of this is the rare genetic disease *alkaptonuria,* in which a single enzyme (homogentisic acid oxidase) is defective. Since the pathway cannot continue beyond homogentisic acid, this substance accumulates, binds to collagen

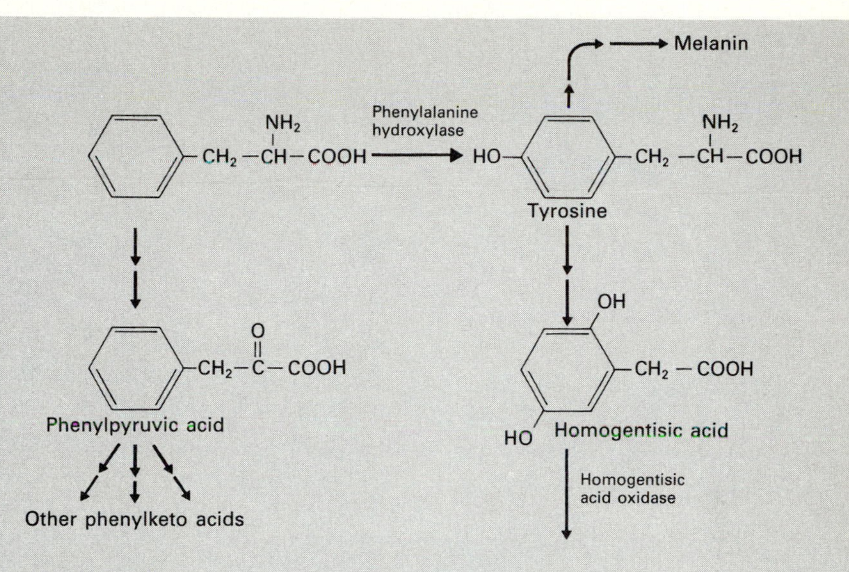

10–9
Outline of the pathways involved in tyrosine and phenylalanine metabolism in human beings. The heavy black line indicates a block in the metabolic pathway due to a defective enzyme. Substances synthesized prior to the block in the pathway accumulate. A block at the arrow produces alkaptonuria, a genetic disease characterized by accumulation of homogentisic acid in the body.

(a protein of connective tissue) and can cause severe arthritis in the joints. Eventually the excess homogentisic acid is excreted in the urine. When the urine is exposed to light, the homogentisic acid turns black. This is one of the symptoms and means of detection of the disease. (Homogentisic acid is also known as "alkapton," hence the name of the disease, "alkaptonuria.")

Chapter 11 Nucleic Acids

11–1 INTRODUCTION: TYPES OF NUCLEIC ACIDS

Why are proteins so distinctive? To answer this question, attention must be given to the *nucleic acids*. Either directly or indirectly, *nucleic acids determine protein structure*. Today, one of the most exciting areas of biological research centers around trying to discover just *how* nucleic acids perform this protein-building function.

Like proteins, nucleic acids are large molecules with high molecular weights. They are, perhaps, the most fascinating of all macromolecules. Nucleic acids are found in all living organisms from viruses to man. They received their name because of their discovery in the nuclei of white blood cells and fish sperm by Miescher in 1869. However, it is now well established that nucleic acids occur outside of the cell nucleus as well.

Nucleic acids serve two main functions. First, they are a sort of molecular "paper" upon which the blueprint for the construction of a new individual is written. This, essentially, is the function of *deoxyribonucleic acid,* or DNA.

But nucleic acids are more than carriers of the hereditary message. They also serve to put this message into action. This active phase is carried out by *ribonucleic acid,* or RNA.

One of the differences between DNA and RNA is the sugar portion of the molecule. DNA contains deoxyribose, whereas RNA contains ribose.

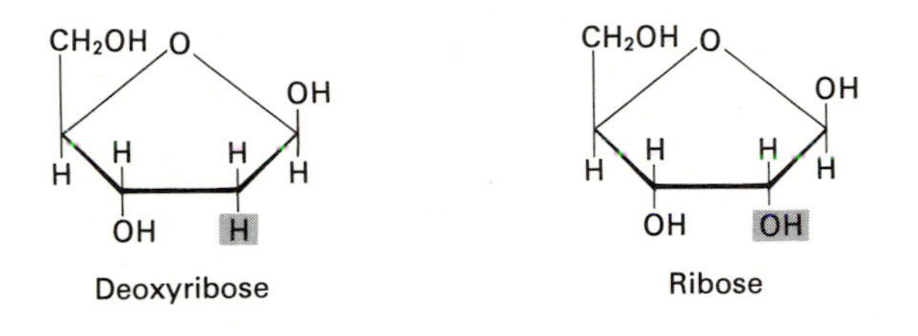

Nucleic acids are generally found within the cell combined with protein to form *nucleoproteins.* In cells of higher plants and animals, complexes of DNA and protein are found within the cell nucleus as rod-shaped bodies called *chromosomes.* Where there is no definite nuclear area, as in bacterial cells, nucleoprotein is scrattered throughout the entire cell. RNA, depending upon its specific molecular type, is found both inside and outside of the cell nucleus, and again, depending upon its specific molecular type, may be found in nucleoprotein complexes (ribosomes). In cells without nuclei, RNA and RNA-protein complexes may be found anywhere within the cell. Before discussing the role of nucleic acids in the process of protein synthesis, let us first consider their structure.

11–2 THE CHEMICAL COMPOSITION OF NUCLEIC ACIDS

Nucleic acids are long-chain molecules composed of units called *nucleotides.* The differences in the nature of individual nucleic acid molecules arise from the order in which these nucleotides are arranged. In this sense, nucleotides are analogous to the amino acids of a polypeptide chain.

Each nucleotide unit is composed of three molecular parts. First, there is a *pentose,* or five-carbon sugar. In RNA, this sugar is *ribose.* In DNA, the pentose sugar has one less oxygen atom, and is thus called *deoxyribose.* Second, there is a *phosphate group* (PO_4). Finally, there is present one of five *nitrogenous* (nitrogen-containing)

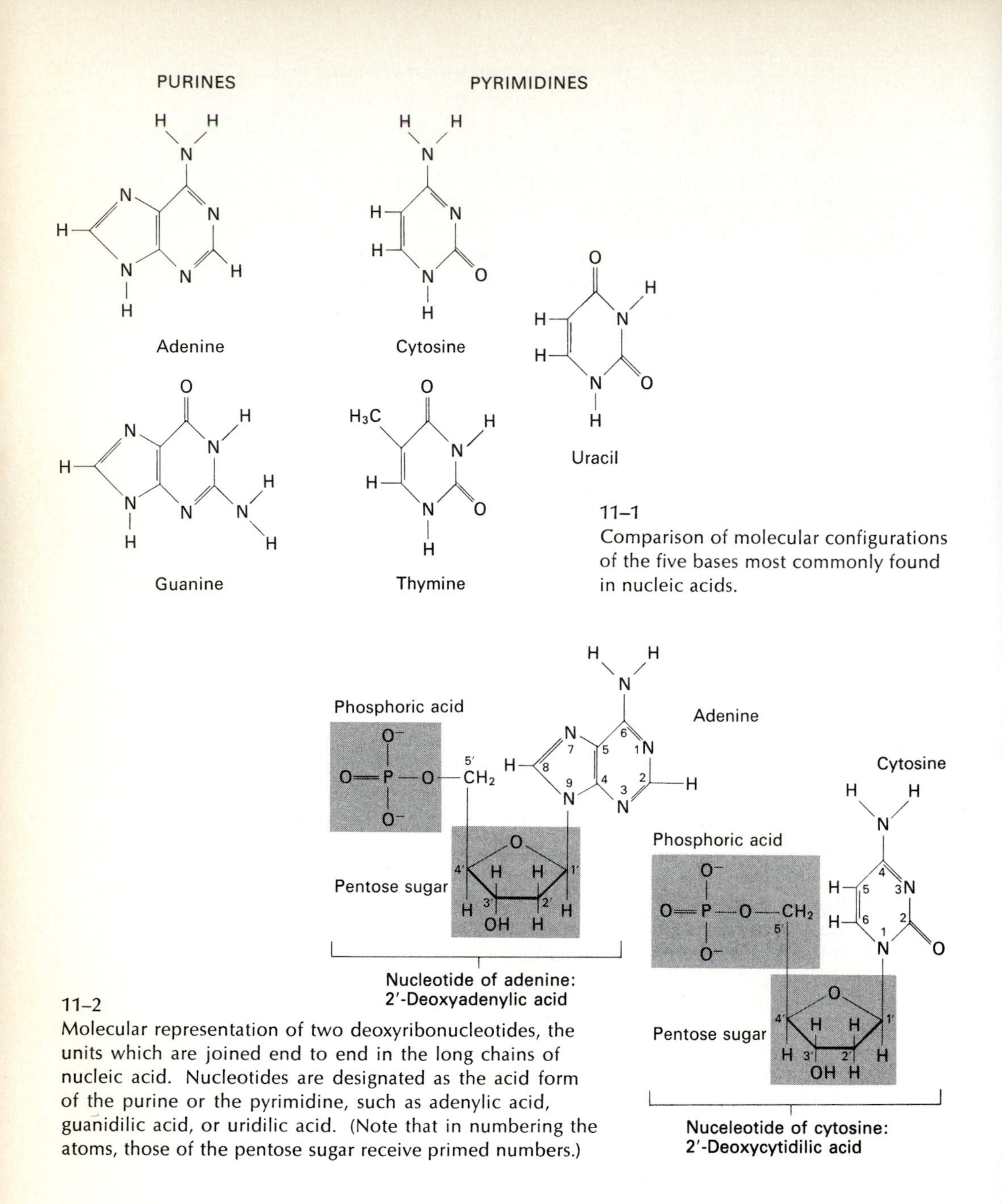

11–1
Comparison of molecular configurations of the five bases most commonly found in nucleic acids.

11–2
Molecular representation of two deoxyribonucleotides, the units which are joined end to end in the long chains of nucleic acid. Nucleotides are designated as the acid form of the purine or the pyrimidine, such as adenylic acid, guanidilic acid, or uridilic acid. (Note that in numbering the atoms, those of the pentose sugar receive primed numbers.)

bases. A nucleotide unit can be diagrammatically represented as:

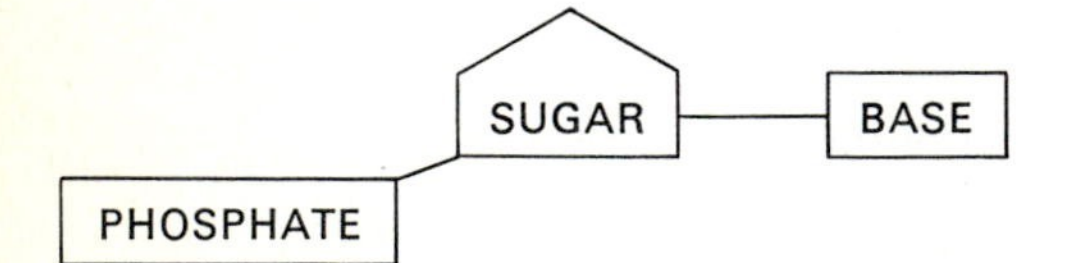

The five bases most commonly found in nucleic acids are listed in Table 11–1.

TABLE 11–1 Five organic bases found in nucleic acids*

Name	Most commonly found in	Symbol
Adenine	DNA and RNA	A
Guanine	DNA and RNA	G
Cytosine	DNA and RNA	C
Thymine	DNA only	T
Uracil	RNA only	U

* Other bases besides those listed here can be found in certain naturally occurring and artificially produced nucleic acids. However, these five are by far the most frequently encountered and for this reason will be those with which we shall deal in discussing the chemical structure of nucleic acids.

These bases belong to two general groups of compounds, *purines* and *pyrimidines*. Purines are double-ring nitrogenous bases, whereas pyrimidines are single ring forms. Adenine and guanine are purines. Cytosine, thymine, and uracil are pyrimidines. The structural formulas of these five compounds are shown in Fig. 11–1; see also Fig. 11–2.

A quick glance at their molecular formulas will indicate why the purines and pyrimidines are called bases. The

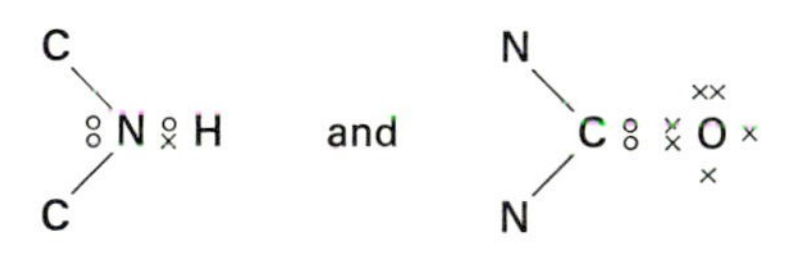

regions of the molecules all possess unshared electron pairs, symbolized by the two open dots to the left of the N in the first group and, in the second group, by the four crosses of the O which are not shared by the C. These electron pairs serve to attract protons and the various nitrogenous bases act as proton acceptors. It is for this reason they are referred to as bases.

Single nucleotides can be joined together end to end to form long, single-stranded *polynucleotides*. Linking of two nucleotides occurs between the phosphate group of one unit and the sugar group of another. This specific bonding is shown in Fig. 11–3. From the sugar-phosphate backbone thus formed, the individual bases protrude as diagrammed below:

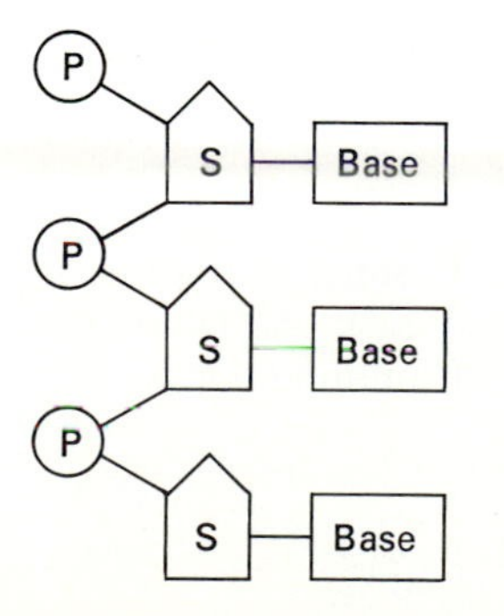

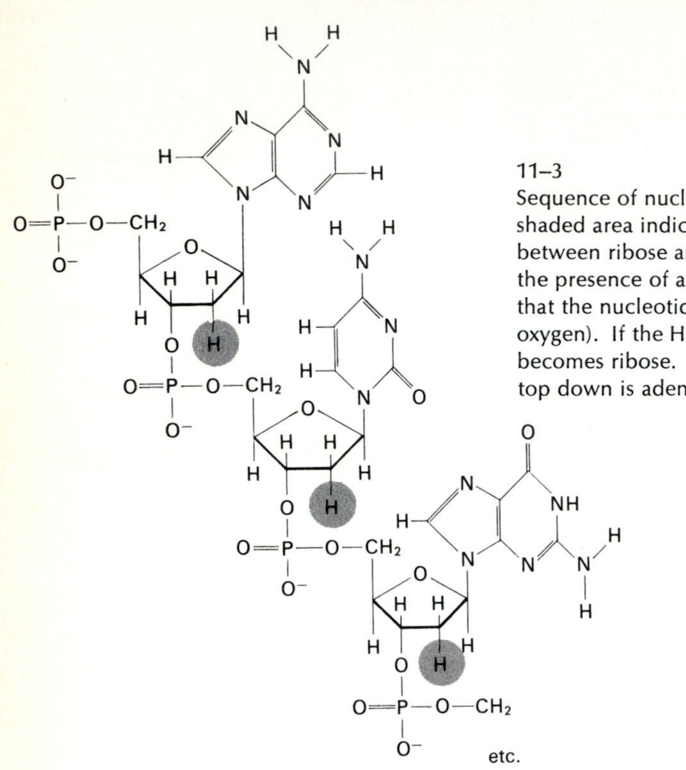

11–3
Sequence of nucleotides in one strand of DNA. The shaded area indicates the region of difference between ribose and deoxyribose. In this diagram, the presence of a single H in these regions indicates that the nucleotide contains deoxyribose (without oxygen). If the H is replaced by an —OH, the sugar becomes ribose. The sequence of bases from the top down is adenine, cytosine, and guanine.

Regardless of the nucleotides involved, the sugar-phosphate backbone* for each type of nucleic acid is always the same. Like proteins, nucleic acids attain specificity by the precise ordering of several molecular units in a very long chain. The major variation, therefore, is in the sequence of purine and pyrimidine bases attached to this sugar-phosphate backbone. Whereas the number of types of building blocks for proteins (amino acids) is about twenty, the number for nucleic acids is generally limited to four. Despite this, an almost infinite variety of nucleic acids can be built by simple rearrangement of these parts.

11–3 THE STRUCTURE OF DNA

For many years, the structure of DNA was a mystery. Certain scattered bits of information, however, led to the formulation of a general picture. Electron microscope photographs of purified DNA from

* The "backbone" refers to the sugar-phosphate chain from which the various bases extend.

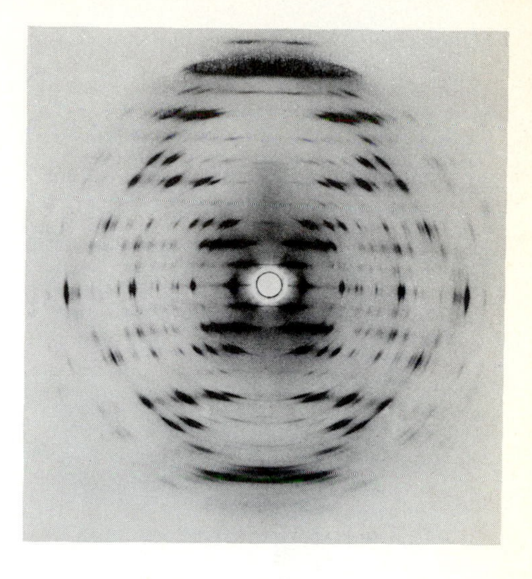

11–4
The x-ray diffraction pattern of DNA obtained from the work of M. H. F. Wilkins. The circular arrangements of dots at ever wider diameters indicate the spiral nature of a long DNA chain. (Courtesy M. H. F. Wilkins.)

viruses indicated that the molecule was long and threadlike. Its thickness of 20 angstroms suggested that the whole molecule might be only 10 to 12 atoms in diameter.

More important information, however, came from x-ray diffraction studies and chemical tests. In 1953, using x-ray diffraction patterns obtained by M. H. F. Wilkins, J. D. Watson and F. H. C. Crick proposed a model to explain the structure of DNA. In this model, each DNA molecule was pictured as consisting of two long chains of nucleotides. These chains twisted in opposite directions so that they fitted together, much as two pieces of rope might be entwined. This double helix formed the backbone of the Watson-Crick model of DNA.

The outside of each polynucleotide strand was envisioned as being composed of a sugar-phosphate backbone, with the purine and pyrimidine bases pointing inward. The bases from one strand face the bases from the other strand to form the central core of the molecule. Thus the entire DNA molecule resembles a spiral staircase. The bannisters represent the sugar-phosphate backbone, while the individual steps represent the bases extending toward the center from each sugar-phosphate strand. A diagrammatic representation of the Watson and Crick DNA molecule is shown in Fig. 11–5.

How are the two strands held together? Certainly two pieces of entwined rope will uncoil if their ends are not tied. Watson and Crick suggested that hydrogen bonds are formed between the two bases which lie across from each other in the double helix. Hydrogen bonds connect each nucleotide of one strand to the nucleotide in the opposite sugar-phosphate strand. All the way down the molecule, each base forms hydrogen bonds with the base directly across from it.

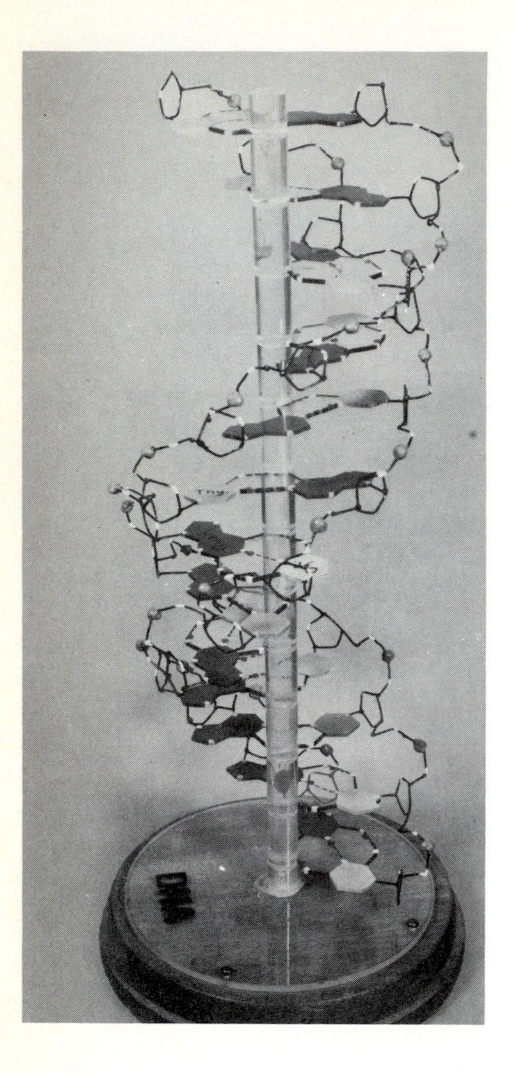

11–5
The Watson–Crick model of DNA structure. The small spheres represent phosphate groups; the open pentagons represent deoxyribose. The solid planar structures represent the purine and pyrimidine bases. (Photo courtesy of Dr. Donald M. Reynolds.)

Is there any pattern to the pairing of the bases? Or will any base on one strand bond with any base on the other? For the answer, Watson and Crick again turned to x-ray diffraction studies. The data showed that the diameter of the molecule was constant throughout its entire length. This constancy could be achieved only if each purine lies across from a pyrimidine. The reason for this is primarily a matter of geometry. Purines are larger molecules than pyrimidines (Fig. 11–1). If two purines were directly opposite each other in a double helix, hydrogen bonding could occur only with difficulty, if at all. The dimensions of these bases would bring the appropriate bonding groups too close together. Were they to remain the proper distance apart for hydrogen bonding to occur, the DNA molecule would show a larger diameter at this point.

TABLE 11–2 Data on purine—pyrimidine composition in some representative DNA's in molar ratios (total = 4.00)*

Source of DNA	Adenine	Thymine	Guanine	Cytosine
Calf thymus cells	1.13	1.11	0.86	0.85
Rat bone marrow	1.15	1.14	0.86	0.82
Gypsy moth	0.84	0.80	1.22	1.33
Silkworm virus	1.17	1.12	0.90	0.81
Bull sperm	1.15	1.09	0.89	0.83

* Modified from data given in G. R. Wyatt, *Chemistry and Physiology of the Nucleus.* New York: Academic Press, Inc., 1952.

Chemical analysis strongly supported the Watson-Crick suggestion that only a union between a purine and a pyrimidine occurs, never one between two purines or two pyrimidines. Evidence like that recorded in Table 11–2 showed that the percentage of adenine was quite similar to the percentage of the thymine. The percentage of guanine, although it might differ from that of adenine or thymine, was close to the percentage of cytosine. From these experimental data it was easy to deduce that adenine will form hydrogen bonds only with thymine and guanine only with cytosine in the DNA molecule. Watson and Crick incorporated this idea into their hypothesized model. The specific pattern of hydrogen bonding between purines and pyrimidines is shown in Fig. 11–6.

Thus, base pairing between opposite strands of any DNA molecule is fixed. In other words, adenine is always paired with thymine, and cytosine is always paired with guanine. It follows, therefore, that whenever one strand contains adenine, the other will contain thymine, and where one contains cytosine, the other will contain guanine. If the sequence of bases on one strand is

A T G G C A A T C,

the sequence of bases on the opposite, or complementary, strand of the molecule will be

T A C C G T T A G.

It can be inferred from this that the sequence of bases on one strand determines the sequence of bases on the other (Fig. 11–7).

The Watson-Crick conception of DNA, like most scientific models, is simply an attempt to explain a mass of experimental data. At least one form of DNA has been found which appears to be single stranded, but this does not invalidate the Watson-Crick hypothesis, since most of the experimental evidence accumulated since 1953 has given it unqualified support. The hypothesis not only explained all of the data available at the time it was proposed, but has also led to a number of new predictions and further discoveries.

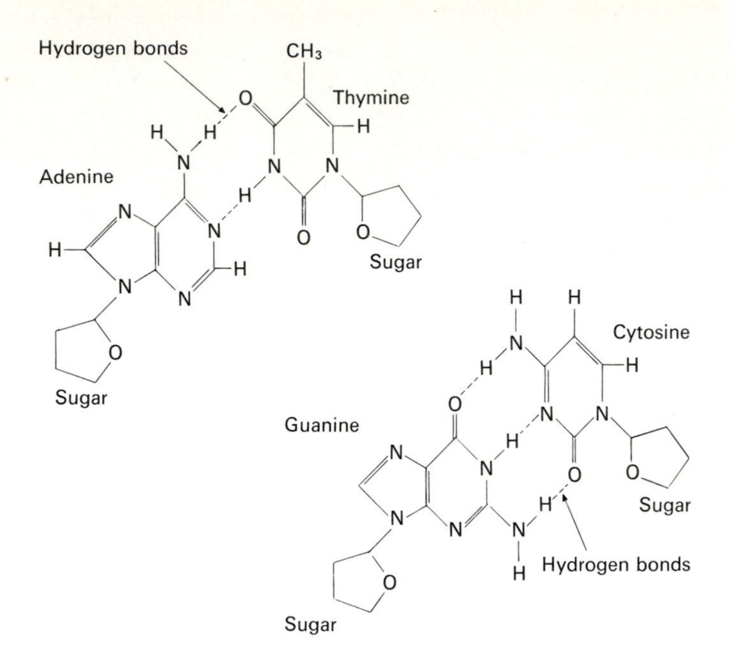

11–6
The structural formulas of adenine, thymine, cytosine, and guanine above show the specific hydrogen-bond formation between purines and pyrimidines. In DNA molecules, adenine always pairs with thymine and guanine pairs with cytosine.

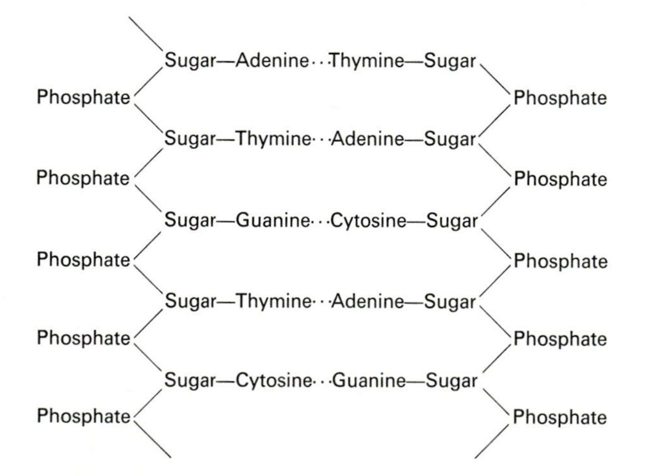

11–7
The Watson-Crick hypothesis suggests that a single DNA molecule is composed of two complementary strands. Each of these is composed of a sugar-phosphate backbone with the bases pairing across the strands.

11–4 THE REPLICATION OF DNA

One of the most satisfying aspects of the Watson-Crick model of DNA structure is that it readily provides a simple means to explain how the molecule might *replicate,* i.e., build exact copies of itself. Indeed, DNA can replicate in either living or nonliving systems. In a living organism, it is the ability of DNA to replicate which allows daughter cells to contain the same genetic message as the original parent cell. The reproduction of almost all forms of life can ultimately be traced to the replication of DNA.

With the Watson-Crick model as a starting point, one early hypothesis explained DNA replication by proposing that the two strands of the DNA molecule separate or "unzip," with each separated strand then acting as a template for the formation of a new, complementary strand (see Fig. 11–8). The nucleotide bases composing DNA attach to their respective complementary partners available in the surrounding medium. In this way, a second strand is built up continuously, as the molecule "unzips," alongside each of the complementary strands of the original "parent" DNA molecule. Each new molecule is thus an exact replica of the original.

This model of the means of DNA replication has received a great deal of experimental evidence to support it. Nevertheless, its accuracy has been seriously challenged by the Nobel laureate Arthur Kornberg (who, with a team of Stanford University biologists, succeeded in synthesizing biologically active molecules of DNA).

To understand the replacement of the preceding explanation of DNA replication with another (hereafter referred to as the Kornberg-Okazaki hypothesis), it must be pointed out that there are two ends of a single-stranded length of DNA—one known as the 3′ end (referring to carbon number 3 on the deoxyribose portion of the nucleotide), the other known as the 5′ end (referring to carbon number 5 of the deoxyribose (see Figs. 11–2 and 11–8)). When two complementary strands are twisted together to form the familiar DNA double helix, they lie together "head-to-tail," with the 3′ end of one single strand opposite the 5′ end of the other. Now, if the DNA polymerase works as suggested by the first hypothesis, i.e., that both strands of the original DNA molecule are working together and in the same manner to bring about replication, then the enzyme must be able to work just as well going from a 3′ to a 5′ end as from a 5′ to a 3′ end. In other words, DNA polymerase must be able to work both "forward and backward." However, many experiments reveal that DNA polymerase cannot do so—that it can only work from the 3′ end to the 5′ end of the strand it is copying, and not in the other direction. The existence of another enzyme which works in the opposite direction to synthesize the other strand can, of course, be hypothesized. Thus far, however, exhaustive searching has failed to detect such an enzyme.

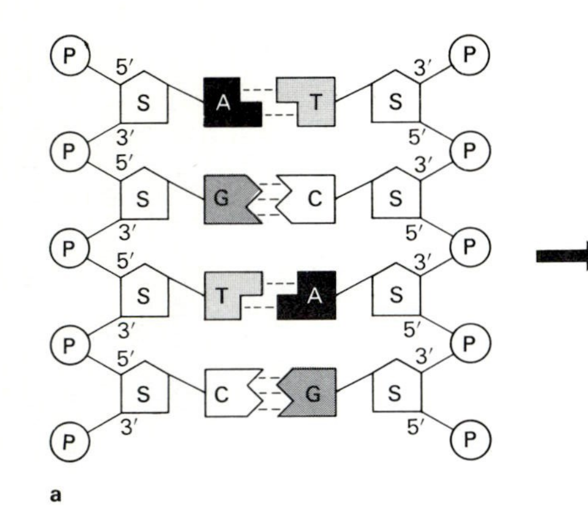

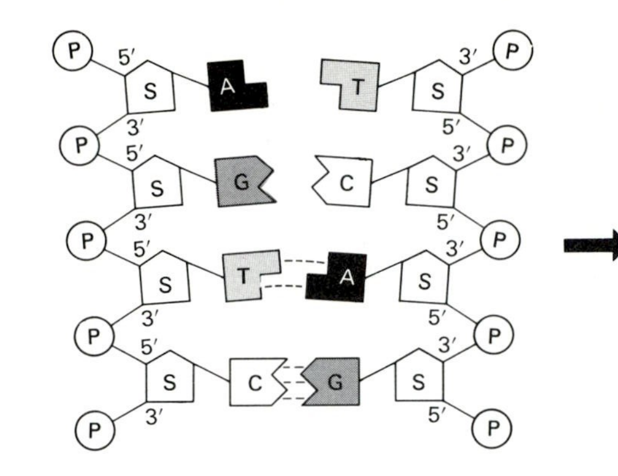

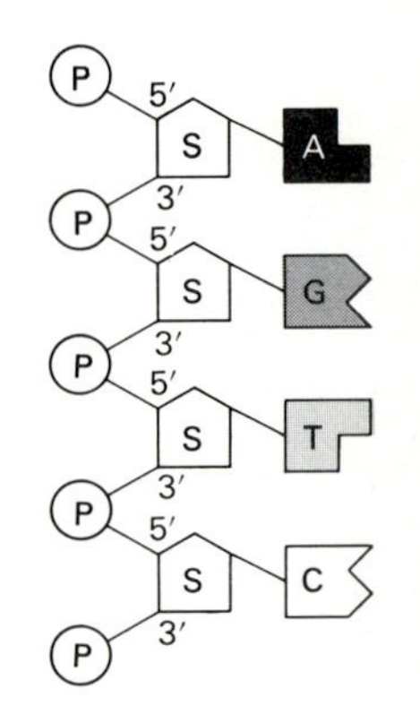

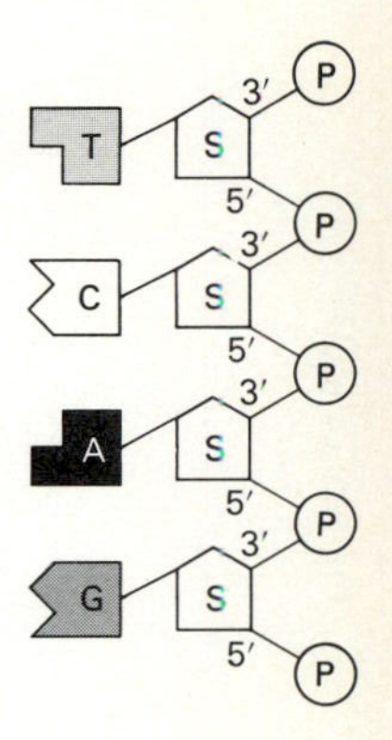

a

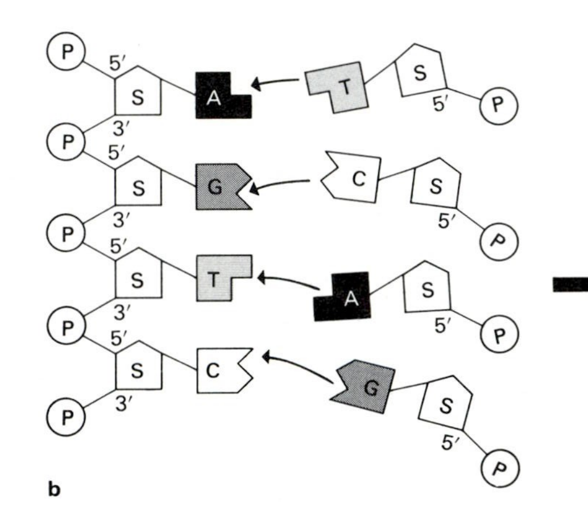

b

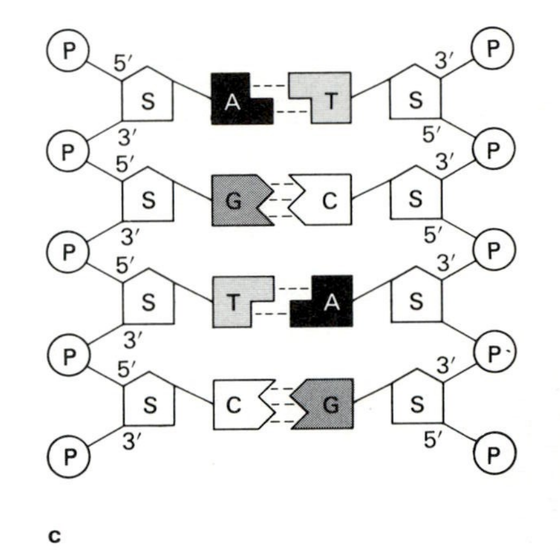

c

◀ 11-8
An early version of DNA replication, based on the Watson–Crick model. Only a short segment of four nucleotide units in each complementary strand is represented. For clarity, no attempt has been made to show the three-dimensional, helical structure of the DNA molecule. The primary enzyme involved in the DNA replication process is DNA polymerase. (a) The original DNA molecule begins to unwind at one end, causing separation of the two helical strands. This separation can be brought about in the test tube by decreasing the concentration of various ions or by increasing the temperature. The unwinding process is a result of the breaking of hydrogen bonds between base pairs on the complementary strands of each DNA molecule. Like the two sides of a zipper, the two strands separate and pull apart. The dotted lines indicate hydrogen bonds. (b) Each strand of the original DNA molecule forms hydrogen bonds between its own unpaired bases and complementary bases available in the medium. Here, for example, a loose thymine unit with its attached ribose sugar and 5′ phosphate moves toward an unpaired adenine unit. Similar moves are made by loose adenine, guanine, and cytosine units toward their unpaired complementary bases. (c) Final stage of complete DNA replication. The sugar-phosphate units on the newly formed complementary strand join together in such a way that each phosphate group unites two sugars. This occurs when DNA polymerase causes a bond to be formed between the phosphate on the 5′ carbon of the pentose of one nucleotide and the 3′ carbon of the adjacent nucleotide's pentose. Since the other strand of the original DNA molecule has undergone the same process, the result is two identical copies of the original molecule. This replication is followed by the division of the chromosomes within the cell nucleus and eventually by division of the cell itself.

The older "unzipping" hypothesis cannot be completely discarded, for there is much to support it. For example, when a zipper unzips, a Y is formed, with the unzipped portion represented by the stem of the Y. All experimental evidence indicates that the new daughter strands are synthesized simultaneously at the point where the Y branches out, i.e., where it is first beginning to open. To account for this fact, and also to explain how simultaneous synthesis can occur on both strands using an enzyme which can only work in a 3′ to 5′ direction, is a difficult feat.

Difficult, but not impossible. Figure 11–9 shows how DNA replication is now hypothesized to occur. Note that in one branch of the replicating molecule, the DNA polymerase moves from the 3′ to the 5′ end of the template strand, putting the new strand together in one continuous string, just as the Watson-Crick model predicts. However, at the same time, a complement to the other strand is being built *in the opposite direction*. In other words, the DNA polymerase is still working in a 3′ to 5′ direction.

This hypothesized mode of reproduction necessarily leads to two testable predictions. First, since DNA polymerase works only in the apex region of the Y, it must be predicted that this second strand is synthesized in short, disconnected segments. A group working under

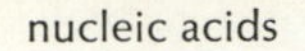
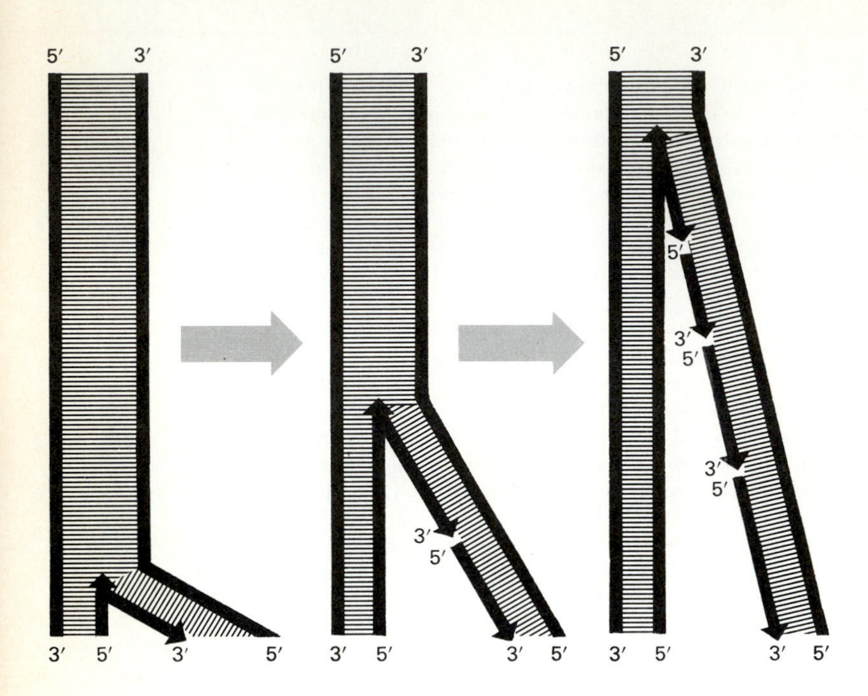

11–9
DNA replication as pictured by the Kornberg-Okazaki hypothesis. One of the parent strands is copied "backwards" in short segments 1000 to 2000 nucleotide units long. Thus the DNA polymerase need travel only in the 3′ to 5′ direction. A joining enzyme later links these short segments together to form a continuous strand.

Dr. Reiji Okazaki at Nagoyo University in Japan showed that in bacteria, if a short radioactive pulse is used to label the most recently made segment of DNA, a large proportion of the radioactive label is indeed found in short, unconnected chains. In fact, nearly all the radioactive label turned up in short DNA sections of one to two thousand nucleotide units in lengths. This result led Okazaki to go so far as to suggest that both daughter strands, and not just the one synthesized "backwards," are made discontinuously.

If one or both strands of DNA are synthesized in segments, rather than in one or more complete strands, then there must be an enzyme responsible for catalyzing the reaction that joins these disconnected segments together. In his attempts to synthesize a viral DNA, Kornberg had tried without success to use DNA polymerase to do the complete job. Then a new enzyme—one which would join short segments of DNA together—was discovered.* This enzyme provided

* This "joining" enzyme, along with DNA polymerase, was used by Kornberg to synthesize his biologically active DNA.

the vital ingredient for survival of the Kornberg-Okazaki hypothesis of discontinuous DNA replication.

Further support for the Kornberg-Okazaki hypothesis comes from the latter's 1968 work with a bacteriophage (a virus which attacks bacteria) producing a "joining" enzyme which is inactive at high temperatures. By warming the bacteriophage, Okazaki was able to show a sharp and rapid increase in the normal number of labeled short lengths of DNA—a result which would be immediately predicted by the Kornberg-Okazaki hypothesis of DNA replication.

While there may be a great deal of evidence to support the preceding description of DNA replication, there is far less indicating that the same procedure occurs in living matter. It is quite difficult to study such delicate chemical reactions *in vivo* (within a living system) without at the same time causing the cessation of life. Much more work will have to be done before the exact mechanism of DNA replication becomes clear. Until a better model leading to still more accurate predictions is proposed, however, the present Kornberg-Okazaki hypothesis based on the Watson-Crick model offers the best available explanation for the fascinating phenomenon of molecular reproduction.

11–5 THE STRUCTURE OF RNA

The role of ribonucleic acid (RNA) in living organisms is more varied than that of DNA. Some viruses, for example, contain no DNA whatsoever. Viruses are composed almost exclusively of nucleic acid (either DNA or RNA) enclosed by a protein coat. Having no other structural parts, viruses can reproduce themselves only by using the cellular machinery of higher plant or animal cells. A single virus can reproduce several hundred copies of itself within twenty minutes after invading a cell. This process destroys the cell and is the reason that viral infections can be so harmful to organisms. In viruses such as the tobacco mosaic virus (TMV) shown in Fig. 11–10, RNA performs the functions normally handled by DNA.

However, in the cells of higher plants and animals, where DNA is present, RNA plays little or no part in the actual transmission of genetic information between generations. Instead, *RNA serves as the intermediate agent by which protein synthesis is accomplished.*

Present ideas concerning the molecular configuration of RNA molecules are much less clear than the Watson-Crick model of DNA discussed above. There are good reasons for this. RNA is far more difficult to obtain in pure crystalline form than DNA. Hence, RNA is less easily studied by x-ray diffraction techniques. As a result, the type of information which this technique can provide has been lacking for RNA, at least until fairly recently. In addition, RNA occurs in at least three forms. Each of these forms has a different structure and

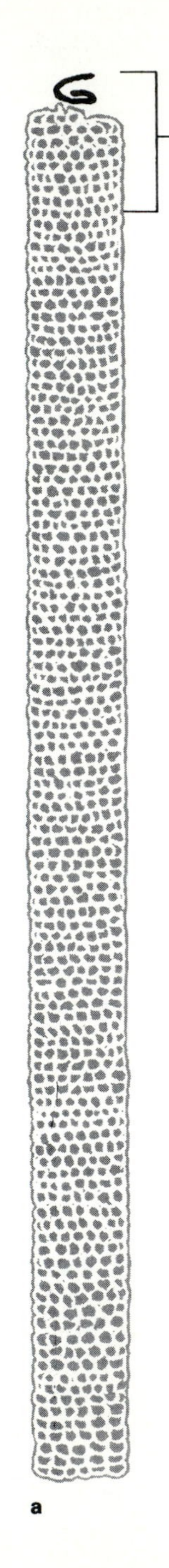

a

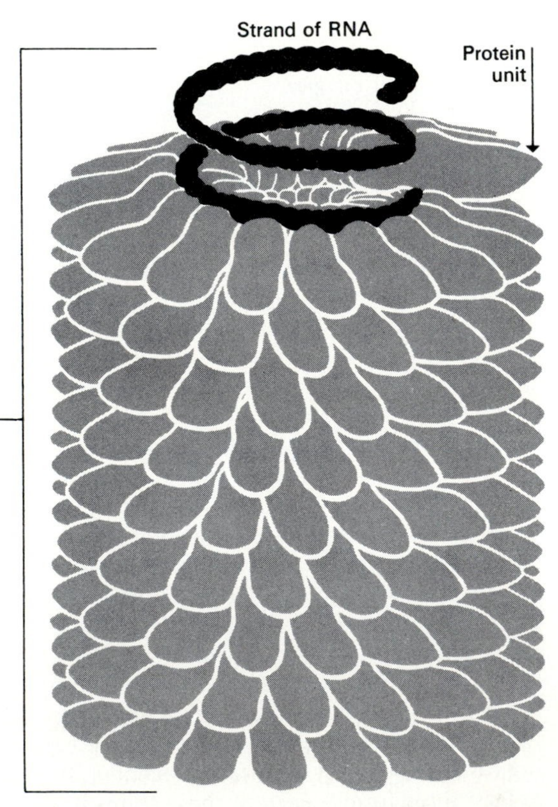

b

11–10
Diagrammatic representation of the structure of the tobacco mosaic virus (TMV). (a) The complete TMV has 2200 protein units. (b) Its single coiled strand of RNA is here seen emerging from the stack of spirally arranged protein units in which it is embedded. All of the protein units in this virus are identical. Each unit is composed of a sequence of 158 amino acids. The precise order of these amino acids has now been determined. (Adapted from W. M. Stanley and E. K. Valens, *Viruses and the Nature of Life*. New York: E. P. Dutton and Co., 1961.)

function. The three forms of RNA recognized today are *transfer* RNA, *ribosomal* RNA, and *messenger* RNA. Structural differences between these forms are not due to differences in the nucleotides involved, but rather are those of molecular weight and configuration.

Transfer RNA (symbolized *t* RNA). Transfer RNA, often called soluble RNA, is the smallest type of RNA. Each transfer RNA molecule contains only 70 to 80 nucleotides. Transfer RNA is the only type of RNA for which a fairly definite molecular structure has been determined.

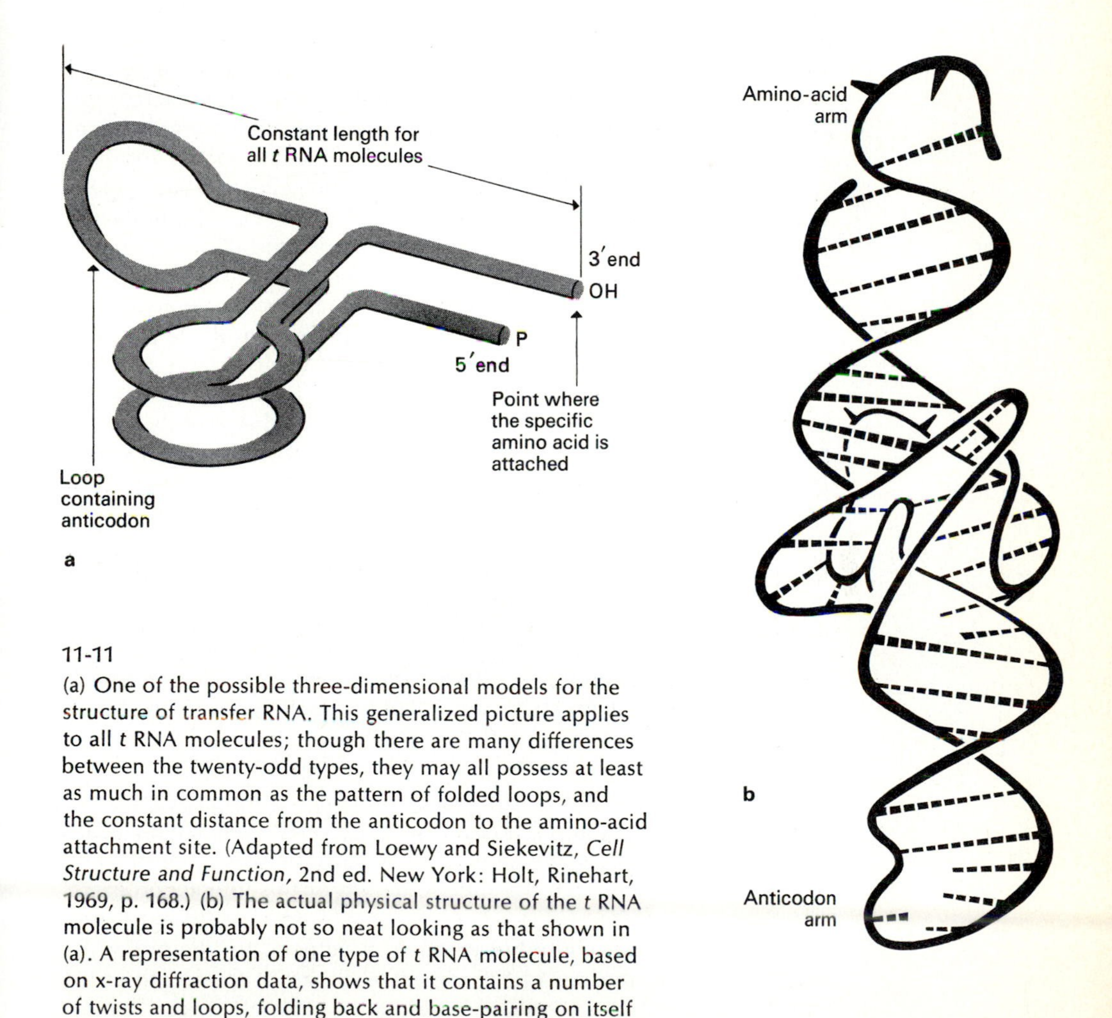

11-11

(a) One of the possible three-dimensional models for the structure of transfer RNA. This generalized picture applies to all *t* RNA molecules; though there are many differences between the twenty-odd types, they may all possess at least as much in common as the pattern of folded loops, and the constant distance from the anticodon to the amino-acid attachment site. (Adapted from Loewy and Siekevitz, *Cell Structure and Function,* 2nd ed. New York: Holt, Rinehart, 1969, p. 168.) (b) The actual physical structure of the *t* RNA molecule is probably not so neat looking as that shown in (a). A representation of one type of *t* RNA molecule, based on x-ray diffraction data, shows that it contains a number of twists and loops, folding back and base-pairing on itself in many places. For purposes of illustrating specific kinds of *t* RNA (see Fig. 11–12), however, the more schematic model shown in (a) will be used.

Transfer RNA picks up individual amino acids in the cell and carries them to the sites of protein synthesis. Since each RNA molecule will pick up only one type of amino acid, there are at least 20 molecular variations of transfer RNA, one for each type of amino acid. Each has a slightly different sequence of bases. This enables each transfer RNA molecule to unite with one specific type of amino acid.

In December, 1964, the precise sequence of the 77 nucleotides of the transfer RNA coding for the amino acid alanine was worked out by a team of Cornell University scientists headed by Dr. Robert Holley. Holley's work, published in 1965, gave precise data only on the sequence of nucleotides in one transfer RNA molecule. At that time there was little evidence about the three-dimensional structure and thus about how the molecule could actually work. In recent years, however, x-ray diffraction studies and even electron microscopy have revealed that the *t* RNA molecule has generally constant shape, although each of the twenty different types (one specifically for each amino acid) have recognizable differences. The generalizable shape is shown in the three-dimensional sketch in Fig. 11–11. The molecule tends to assume this shape as the thermodynamically most stable configuration, with hydrogen bonds forming across the strands between complementary bases. The loop containing the anticodon is where the *t* RNA molecule attaches to the messenger RNA; the anticodon is a triplet complementary to a specific triplet on the *m* RNA, and thus specifying a particular location for its amino acid along the length of the messenger. The —OH, or amino acid arm of the *t* RNA molecule, is where the specific amino acid attaches. The distance from the anticodon loop to the amino acid arm appears to be constant for all *t* RNA molecules, ensuring that all amino acids will be brought adjacent to each other when the *t* RNA attaches to the messenger. Schematic representations of a few types of *t* RNA molecules are shown in Fig. 11–12. There are as many types of *t* RNA in the cell as there are amino-acid types to be incorporated into protein.

Ribosomal RNA (symbolized *r* RNA). Ribosomal RNA has a relatively high molecular weight. It is contained in small box-like structures called ribosomes, located in the cytoplasm of the cell. Ribosomes are composed of roughly 70 percent RNA and 30 percent protein. The ribosomes function as centers of protein synthesis. It is to the ribosomes that the transfer RNA molecules carry their amino acids. During this process, some interaction occurs between transfer and ribosomal RNA, although the precise nature of this interaction is still not clear.

No satisfactory picture has yet been developed for the molecular configuration of ribosomal RNA. For at least part of the length of the molecule, the structure appears to be that of a double helix. The rest of the molecule has an unknown shape. Working with this form of RNA presents special problems of technique. It is extremely difficult to separate the RNA of the ribosome from the protein portion. Attempts to crystallize whole ribosomes have met with only partial

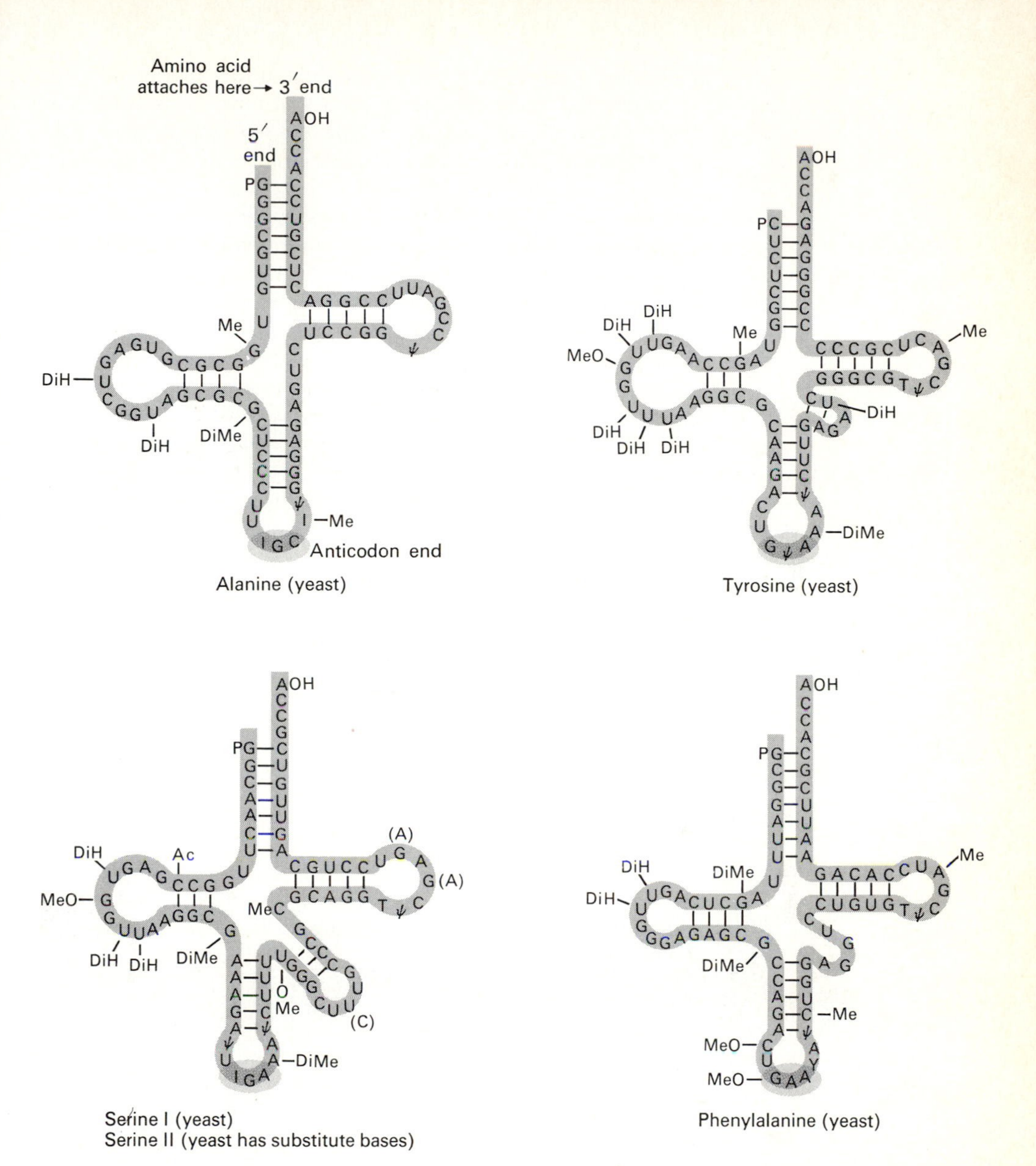

11-12

Several representations of the "cloverleaf" pattern of *t* RNA, showing possible variations in the structure of specific types. Each type of *t* RNA attaches to one type of amino acid at the OH arm shown at the top of each molecule. Some slightly modified bases are incorporated into *t* RNA. These are indicated as DiH-U (dihydroxy-uridine), Me-G (methyl-guanidine), MeO-G (methoxy-guanidine), I (inosine), and ψ (a form similar to uridine). P = phosphate group attached to the 5 end of the molecule (after Holley). (Struther Arnott, "The structure of transfer RNA," *Prog. in Biophys. and Mol. Biol.*, 22 (1971): 181–213; the various *t*- RNA's are diagrammed on pp. 183–185.)

success. For these reasons, a more complete understanding of the molecular configuration of ribosomal RNA will have to wait for the development of more suitable techniques for isolating and crystallizing the pure compound.

Messenger RNA (symbolized *m* RNA). Messenger RNA is the most recently discovered form of RNA. It is intermediate in molecular weight between transfer and ribosomal RNA. In bacteria, messenger RNA is built up and broken down at a rapid rate. Indeed, a molecule of this substance lasts such a short time that there was for a while some doubt as to whether messenger RNA really existed. In the cells of higher organisms messenger RNA may turn over at a much slower rate. It is now agreed that messenger RNA does exist, and that it is a class of molecules heterogeneous in size, reflecting the heterogeneity of protein molecules for which they code.

Messenger RNA carries the genetic code from DNA to the ribosomes, where protein synthesis occurs. As its name implies, messenger RNA carries a message. This message is the genetic "blueprint," or building plan for inherited variation. Messenger RNA transmits the plan in the sequence of its own bases, forming a pattern complementary to that of the DNA which formed it. In the ribosome, the coded message which messenger RNA carries is translated into a specific amino acid sequence. Messenger RNA thus acts as an intermediary between DNA and protein.

11–6 NUCLEIC ACIDS AND PROTEIN SYNTHESIS

The sequence of events involved in producing a specific peptide chain from a genetic code on DNA can be represented as:

DNA → Messenger RNA → Ribosomal RNA → Peptides → Complete protein
↑
Transfer RNA
+
Specific amino acid

The completed protein may be an enzyme. This enzyme acts to control one specific reaction or set of reactions in the cell. Since all enzymes are thought to be produced in this way, the genetic code of DNA ultimately controls the entire metabolic activity of the cell. The experimentally established details of protein synthesis by nucelic acids are described below and illustrated diagrammatically in Fig. 11–13.

By base pairing, DNA forms a molecule of messenger RNA. The nucleotide sequence of this newly formed RNA molecule will be complementary to that of DNA. In other words, if the sequence of bases in the DNA molecule is

A T C C G T G G G A,

then the complementary sequence of bases in the new messenger RNA is

U A G G C A C C C U.

In RNA, the base uracil is substituted for thymine. The completed messenger molecule passes out of the nucleus into the cell cytoplasm where it comes in contact with ribosomes, small cell organelles composed of protein and ribosomal RNA.

Meanwhile, in the cytoplasm another series of reactions is taking place. Before the amino acids can be joined together into a polypeptide chain, each amino acid must be activated by complexing with an energy-rich compound such as ATP. The activated amino acid is at the same time joined to the *t* RNA molecule specific for that amino acid. The bond formed between the amino acid and the *t* RNA contains slightly more energy than that required for the eventual synthesis of the peptide bond between amino acids. The reaction between an amino acid and the appropriate *t* RNA is summarized as follows:

AA (e.g., alanine) + ATP + *t* RNA (specific for alanine) ⟶ AA---*t* RNA + AMP + 2 inorganic phosphates

The amino acid is attached to the long free arm (amino acid arm) of the *t* RNA molecule as shown in Figs. 11–11 and 11–12. The specificity of the enzyme catalyzing the above reaction assures that alanine and only alanine is linked to the *t* RNA bearing the anticodon for this amino acid. There is a different enzyme with a different specificity for each amino acid and its appropriate *t* RNA.

It is the nature of the *t* RNA molecule, not of the attached amino acid, that now determines where the amino acid is to go in a peptide chain. Experiments have been performed, for example, in which the amino acid cysteine is converted chemically to alanine while attached to the specific cysteine *t* RNA molecule. The modified amino acid is incorporated into the protein as if it were cysteine, indicating that the specific structure of the amino acid does not determine the sequence of units in protein synthesis.

The activated amino acid–*t* RNA complex meets the messenger RNA (*m* RNA) molecule at the ribosome. Ribosomes are the protein-synthesizing machinery of the cell. During protein synthesis, ribosomes are attached to *m* RNA and the AA–*t* RNA complex by weak bonds. There are three specific sites on the ribosome. One of these will attach to a portion of the *m* RNA, while a molecule of AA–*t* RNA will associate with each of the other sites (sites P and A). Just which amino acid bearing *t* RNA associates with the ribosome at any given moment is determined by the triplet code of the *m* RNA interacting with the anticodon of the *t* RNA. Figure 11–13(b) makes the relationship between these components clearer.

Ribosomes perform many chemical functions during protein synthesis; one of these is to bring amino acids into the proper orientation so that covalent bonds (the peptide linkage) can be formed between them. The description that follows presents the major events occurring on the ribosome to accomplish the synthesis of a specific sequence of amino acids.

(1) The first step involves the association of two AA–*t* RNA molecules with the *m* RNA and ribosome complex. Specific soluble factors must be present in the cytoplasm to accomplish this binding by weak forces. (See Fig. 11–13(b).)

(2) An enzyme on the ribosome catalyzes the formation of a peptide linkage between the two amino acids. One molecule of *t* RNA now has two amino acids attached to it. The other has none; it is "empty."

(3) The *m* RNA moves one step across the ribosome. In the process, the "empty" *t* RNA is shoved off the ribosome, and the other *t* RNA bearing the two amino acids is moved with the *m* RNA to occupy the site (site P) vacated by the "empty" *t* RNA. This translocation reaction requires certain soluble factors (e.g., G factor) and an energy input from the hydrolysis of a molecule of GTP to GDP and P_i.

11-13

Generalized schemes depicting protein synthesis. (a) Messenger RNA is synthesized from a single strand of DNA. The DNA unwinds in an enzymatically controlled reaction and one strand serves as the template from which *m* RNA is synthesized by base pairing. The resulting messenger thus contains a sequence of bases which can be read as a linear series of triplets (each called a "codon"). In the above diagram, which strand of the DNA molecule served as the template for the section of *m* RNA shown? (b) The sequence of steps involved in peptide formation at the ribosome. For the sake of convenience, only a single ribosome is shown here, though several ribosomes are usually attached to any one messenger molecule. The sequence shown here (from 1 to 5) shows the steps involved in the addition of one amino acid, and the termination reaction by which the peptide chain is hydrolyzed away from the ribosome, *t* RNA complex (see text for full discussion).

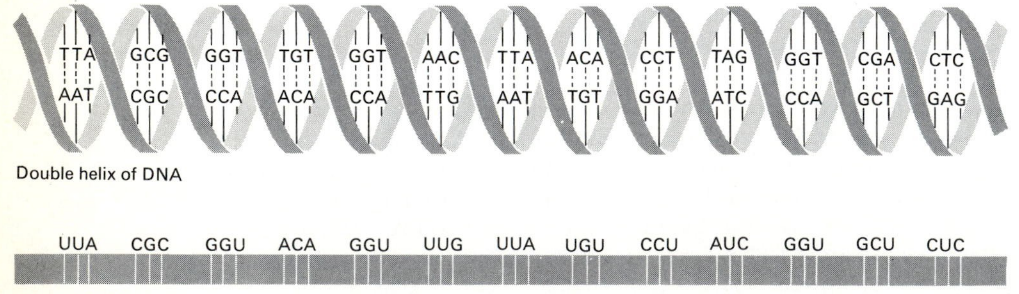

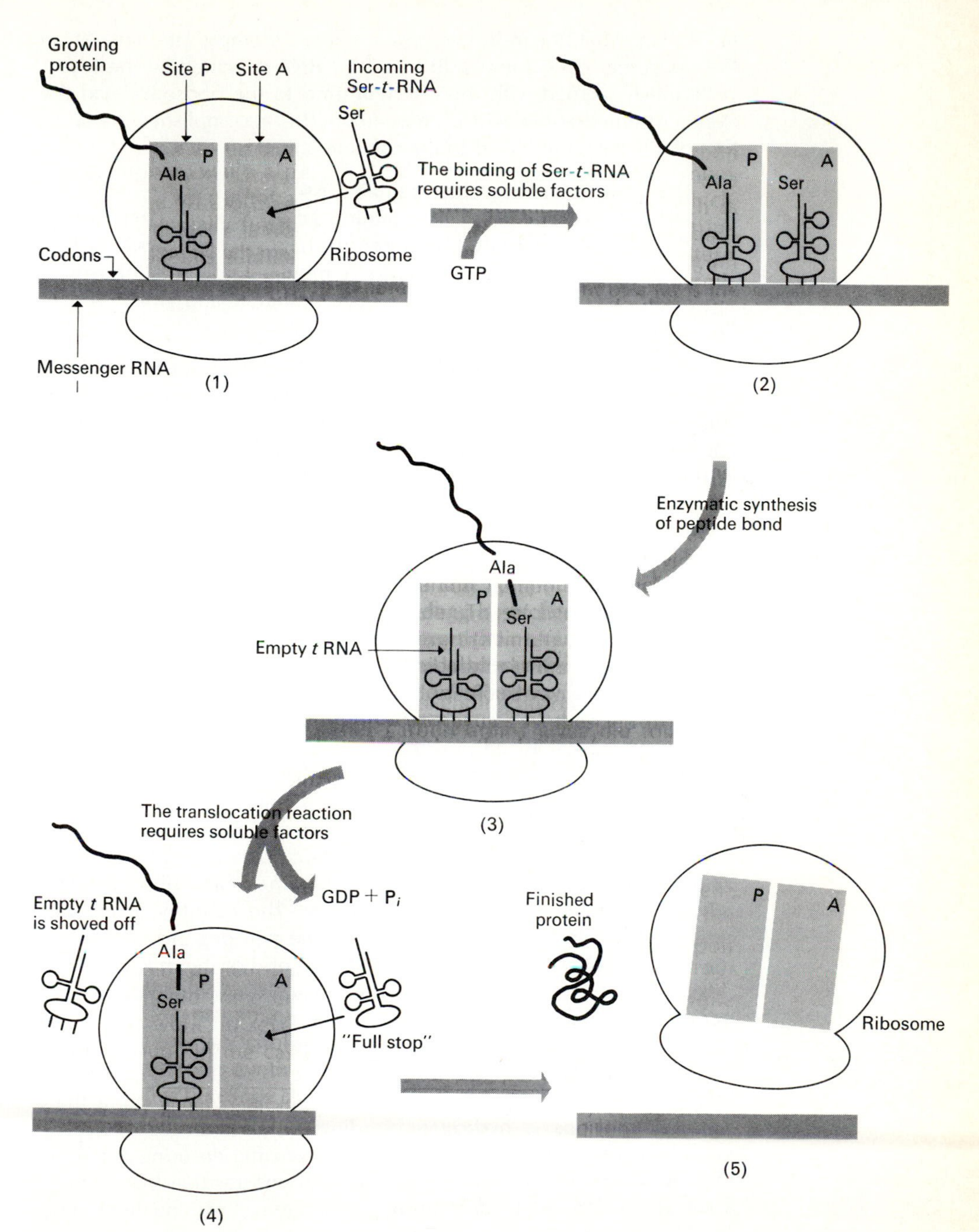
Growing protein
Site P
Site A
Incoming Ser-t-RNA
Ser
P
A
Ala
Codons
Ribosome
Messenger RNA
(1)
The binding of Ser-t-RNA requires soluble factors
GTP
P
A
Ala
Ser
(2)
Enzymatic synthesis of peptide bond
Ala
P
A
Ser
Empty t RNA
(3)
The translocation reaction requires soluble factors
GDP + P$_i$
Empty t RNA is shoved off
Ala
P
A
Ser
"Full stop"
(4)
Finished protein
P
A
Ribosome
(5)

b

(4) As Fig. 11–13(b) indicates, there is still an empty site (site A) on the ribosome. The amino acid bearing *t* RNA specified by the triplet code newly aligned with this site will bind to the ribosome, and the events outlined above will be repeated. This accomplishes the addition of another amino acid to the peptide chain.

The growing peptide chain remains associated with the ribosome through each successively added amino acid. Always attached to a *t* RNA molecule, it is passed back and forth from site A to site P, P to A, A to P, and so on as the *m* RNA moves across the ribosome. Eventually the peptide chain will be terminated and thus become free from the ribosome.

Termination is not a random process, but turns out to be highly controlled by special "full-stop" or termination codons. Recent studies have shown that certain triplets or codons (called "nonsense codons") in the *m* RNA chain automatically bring about termination of the peptide chain at that point (see Fig. 11–13, step 4). The nonsense codons are UAA, UAG and UGA. When the ribosome reaches a nonsense codon, the bond between the final amino acid and the *t* RNA molecule to which it is attached is hydrolyzed. This reaction is mediated by a protein "release factor" which may act as an enzyme, or in some other capacity not clearly understood at present. Thus, the final amino acid is released from its *t* RNA molecule without forming a peptide bond with another amino acid, as is usually the case. The exact chemistry of termination is still being actively investigated. It appears to be built into the genetic message of the DNA as accurately as the position of each amino acid.

Alexander Rich and his co-workers have found that the long molecule of *m* RNA may have more than one ribosome associated with it. Electron microscopy can reveal several ribosomes spaced at intervals along the length of *m* RNA. Such a cluster of ribosomes held together by *m* RNA is called a polysome. Each ribosome of a polysome is involved in protein synthesis. The quantity of growing protein associated with any given ribosome will depend on how far that ribosome has traveled from the starting end of the *m* RNA. Thus, it is clear that each molecule of *m* RNA may serve to generate a number of identical proteins from its coded message.

The events of DNA replication, and protein and RNA synthesis all depend upon the formation of specific, though weak, chemical interactions among the nucleic acids involved. Chief among these weak interactions is hydrogen bonding between specific pairs of complementary bases. These are highly specific, ensuring the accuracy of the processes. Their weak character ensures that the chemical associations formed in determining a sequence will be temporary. In the case of protein synthesis, the weak bonds ensure that the newly created substance can be easily detached from the messenger RNA and the ribosome.

The end product of the above sequence of reactions is a completed protein molecule. In many cases, however, a protein can acquire all of its functional properties (e.g., enzymatic) only when it is closely bound to other proteins. The fully functional protein is then said to be made up of *protein subunits*. Such a protein may be an enzyme, or it may be a structural or transport protein such as collagen or hemoglobin. The primary structure of all proteins is determined at the ribosome by the mechanism outlined above. The coiling of a single peptide chain into the α-helix is determined by intramolecular forces associated with the elements of the peptide linkage. The combination of various coiled proteins into a fully functional protein molecule is determined by the number and location of each amino acid and the interactions that are possible between their side chains (secondary, tertiary, and quaternary structure). (Quaternary structure refers to the interactions of polypeptide subunits within a protein.) Because these intricate conformations of proteins are dependent upon the specification of the amino-acid sequence, we can say that all information about the cell's structure and function resides in the sequence of bases in the DNA molecule.

The reactions described above have been studied in greatest detail in bacteria. To a large extent, a similar sequence of steps is thought to occur in the cells of higher organisms (animal and plant) as well. While there are some notable differences, the similarities confirm the generalization that all life seems to possess a biochemical unity: to be based on very similar biochemical properties.

11–7 THE GENETIC CODE

While it is beyond the scope and aims of this book to go into the details of research on the genetic code, it is fitting, perhaps, at least to define the problem. Certainly, without the genetic code, the story of nucleic acids is incomplete.

We have seen that the sequence of bases on ribosomal RNA, originally obtained via messenger RNA from DNA, determines the sequence of transfer RNA molecules along the ribosomal RNA molecule. This transfer RNA sequence, in turn, establishes the amino acid sequence in the peptide chain being constructed. The problem of the genetic code, however, is even more specific. The question it poses has three parts. First, how many nitrogenous bases are involved in selection of the amino acid-carrying transfer RNA molecules? Second, which of the four bases are involved? Third, in what order are these bases arranged?

The first question was attacked by simple arithmetic. It was obvious that more than one base was involved. With only one base playing a role, only four amino acids could be selected. After all, there are only four kinds of bases in RNA.

TABLE 11–3 Possible genetic-code letter combinations as a function of the length of the code word. (After M. Nirenberg, 1963)

SINGLET CODE (4 WORDS)	DOUBLET CODE (16 WORDS)				TRIPLET CODE (64 WORDS)			
A	AA	AG	AC	AU	AAA	AAG	AAC	AAU
G	GA	GG	GC	GU	AGA	AGG	AGC	AGU
C	CA	CG	CC	CU	ACA	ACG	ACC	ACU
U	UA	UG	UC	UU	AUA	AUG	AUC	AUU
					GAA	GAG	GAC	GAU
					GGA	GGG	GGC	GGU
					GCA	GCG	GCC	GCU
					GUA	GUG	GUC	GUU
					CAA	CAG	CAC	CAU
					CGA	CGG	CGC	CGU
					CCA	CCG	CCC	CCU
					CUA	CUG	CUC	CUU
					UAA	UAG	UAC	UAU
					UGA	UGG	UGC	UGU
					UCA	UCG	UCC	UCU
					UUA	UUG	UUC	UUU

A = adenine, G = guanine, U = uracil, C = cytosine.

Could there be two bases involved? Again, with only two bases, the possible arrangements are 4^2, or 16. This number is not large enough to allow for the selection of the 20 amino acids known to be used in protein synthesis.

The minimum number of bases which could be involved in amino acid selection seemed to be 3. This number gave a possibility for the selection of 4^3, or 64 amino acids, more than enough. Higher numbers than 3 were possible, of course, but seemed unnecessary and thus, it was assumed, less likely.

Work on the genetic code proceeded, therefore, on the assumption that it was a *triplet* code, involving only 3 bases. The experimental evidence thus far strongly indicates that this assumption is correct. Indirectly, through transfer RNA 3 bases are involved in amino acid selection. Table 11–3 gives the coding as known in 1963.

All the possible codons have now been associated with specific amino acids or a role in polypeptide synthesis (see Table 11–4), although some of the codons are assigned with less certainty than others. Table 11–4 gives the triplets of base pairs for each of the 20 most common amino acids. To select the proper triplets or codons for any amino acid, simply read in order the letters appearing to the left, above, and to the right of it. Thus, the codons for glycine (gly)

TABLE 11–4 The genetic code

1ST ↓ 2ND →	U	C	A	G	↓3RD
U	phe	ser	tyr	cys	U
	phe	ser	tyr	cys	C
	leu	ser	NONSENSE (OCHRE)	STOP	A
	leu	ser	NONSENSE (AMBER)	tryp	G
C	leu	pro	his	arg	U
	leu	pro	his	arg	C
	leu	pro	gluN	arg	A
	leu	pro	gluN	arg	G
A	ileu	thr	aspN	ser	U
	ileu	thr	aspN	ser	C
	ileu	thr	lys	arg	A
	met	thr	lys	arg	G
G	val	ala	asp	gly	U
	val	ala	asp	gly	C
	val	ala	glu	gly	A
	val	ala	glu	gly	G

are GGU, GGC, GGA, and GGG, while those for lysine (lys) are AAA and AAG. Some of the codons assigned here are less certain than others, and those suspected of being connected with the beginning or termination of polypeptide chain synthesis are not included.

Quite surprising was the discovery that the genetic code is *degenerate*. This means that there may be more than one triplet of bases which can select the same amino acid. For example, the triplets AAU and AAC both select asparagine.

All the experimentation on the genetic code was made possible by learning to make synthetic polynucleotides. In this way, the sequence and type of bases in the strand could be specified. It then became a matter of seeing which amino acids were selected by the known triplet combinations. We have seen that with 4 bases arranged randomly in triplets, 64 different combinations are possible. Now, all 64 triplets have been linked to the selection of amino acids.

There is an unfortunate tendency to view solution of the genetic code as a major breakthrough marking the end of the road. Solution of the complete code is a major breakthrough, indeed. However, it marks a *beginning,* not an end. The problems of how a cell uses the completed protein at the proper time and place in cellular differentiation, specialization, and organization are still a very long way from being solved. Knowing how its buildings are made tells us little or nothing of the overall plans for a city.

There are, indeed, exciting times ahead.

MECHANISM OF ACTION OF TWO ANTIBIOTICS

Bacteria can reproduce every 20 minutes. It is thus easy to imagine how quickly a bacterial population can grow and how a bacterial infection can be manifested. In order to limit the growth of bacteria when our own defenses have been overcome, we often take antibiotics.

In order for all cells to survive and especially to grow and divide, they must be able to carry on protein synthesis. This involves transcription of *m*RNA from the DNA and subsequent translation of *m*RNA to protein at the ribosomes.

The transcription of *m*RNA from DNA can be prevented by the administration of the antibiotic actinomycin D, which inserts itself into specific sites in the DNA double helix. The enzyme that synthesizes *m*RNA from the DNA template is thereby blocked, and no *m*RNA is synthesized. Therefore no protein synthesis can occur. In clinical use, actinomycin and certain other compounds have been found to have specific effects against various tumors and have been applied in treatment of cancer in human beings.

Protein synthesis can be inhibited also at the site of the ribosome. The tetracyclines Terramycin and Aureomycin prevent normal binding of *m*RNA to ribosomes. In this case, RNA can be synthesized from DNA but it cannot be translated into protein. The tetracyclines are used against typhoid fever, various forms of typhus fever, spotted fever, gonorrheal infection, syphilis, urinary tract infections, and other diseases.

Antibiotics are selective in their antimicrobial action; i.e., for each antibiotic, some organisms are affected and others are not. But the antibiotics are also effective against animal cells, with different antibiotics having different degrees of toxicity to the animals. Actinomycin is highly toxic to animal tissues. Tetracyclines can disturb the natural bacterial populations found in the intestines and can cause gastrointestinal upsets and give rise to secondary infections.

As with nearly all drugs, a risk/benefit decision must be made with each administration: which is greater—the harmful effects or the beneficial effects? Substances are incorporated more quickly into rapidly growing cells and tissues (such as bacteria and tumors) than into slowly growing tissues, and therefore certain drugs will have a greater effect on the rapidly growing tissues where they tend to become concentrated. This may be one of the reasons that certain antibiotics are effective in selectively counteracting the growth of the body's invaders.

Source: White, Handler, and Smith, *Principles of Biochemistry* (Fifth ed.), McGraw-Hill, New York, 1973, page 774; and *Collier's Encyclopedia,* (1971) Vol. 2, pp. 314–320.

Suggested Further Readings

GENERAL

A. Inorganic and Organic Chemistry

Borek, E., *The Atoms Within Us.* New York: Columbia University Press, 1961. This is a simplified introduction to the chemistry of living systems, less detailed than *Matter, Energy, and Life.*

Cheldelin, V. H. and R. W. Newburgh, *The Chemistry of Some Life Processes.* New York: Reinhold, 1964. An introduction to organic chemistry with special reference to biochemical processes. This book assumes more background in chemistry and atomic theory than *Matter, Energy, and Life;* it also treats certain biochemical processes in greater detail.

Panares, R. R., *Energy, Organization, and Life.* Chicago: Educational Methods, Inc., 1967. A companion volume to the preceding, this programed text covers the material treated in Chapters 4 and 5 of *Matter, Energy, and Life.* In addition, it covers such topics as photosynthesis, respiration, and carbohydrate metabolism.

Sackheim, G. I., *Introduction to Chemistry for Biology Students*. Chicago: Educational Methods, Inc., 1966. This programed text deals with atomic structure, chemical bonding, catalysis, ionization, and the basic organic compounds found in living systems. It covers the same range of material treated in Chapters 1, 2, 3, 6, 7, 8, and 9 of *Matter, Energy, and Life*.

White, E. H., *Chemical Background for the Biological Sciences*. Englewood Cliffs, N.J.: Prentice-Hall, 1964. This book goes into more depth and detail than *Matter, Energy, and Life,* especially on topics in organic chemistry. The biological applications of various principles are minimized.

B. Biochemistry

Baldwin, E., *The Nature of Biochemistry*. Cambridge: University Press, 1962. This is a short and highly readable introduction to biochemical problems such as protein structure and function, enzymes, amino acids, carbohydrate and fat metabolism, and the nucleic acids. It assumes knowledge on the level at least of *Matter, Energy, and Life*.

Coult, D. A., *Molecules and Cells*. Boston: Houghton-Mifflin, 1966. This book begins with an introduction to basic chemistry, then proceeds to a discussion of protoplasm and cell structure and function, including cellular metabolism. It concludes with a discussion of genetics and development. It is a very biologically oriented book using biological problems as the basis for biochemical discussions.

Jevons, F. R., *The Biochemical Approach to Life*. New York: Basic Books, 1964. This is a general introduction to cells as centers of chemical activity. It deals with many of the topics discussed in *Matter, Energy, and Life,* such as proteins, enzymes, biological oxidation, vitamins and coenzymes, drugs and chemotherapy, ATP, and biochemical genetics. It also has a valuable chapter on the problems of explanation in biology, particularly from the biochemist's point of view.

Lehninger, A. H., *Biochemistry,* New York: Worth, 1970. This is one of the most recent and comprehensive biochemistry texts available. It can be used for reference, but in general requires considerable background (general college inorganic and organic chemistry) for in-depth understanding.

Light, R., *A Brief Introduction to Biochemistry*. New York: Benjamin, 1968. A brief introduction, like Baldwin's, to biochemical problems such as catalysis, energy relationships, metabolism, information transfer (molecular genetics), and the problems of cellular organization and differentiation treated from a biochemical point of view.

chapter one

For the biological and physical background:

Lehninger, A. L., *How Cells Transform Energy, Scientific American* 205 (September, 1961): 62–73

Wald, George, *Life and Light, Scientific American* 201 (September, 1959):92–108.

Both of these articles deal with the overall problems of matter and energy in relation to fundamental processes in living organisms. They provide a background to illustrate the types of problems which modern, chemically-oriented biology is still trying to solve.

chapter two

Bush, G. L. and A. A. Silvidi, *The Atom: A Simplified Description.* New York: Barnes and Noble, 1961. An introductory description of atomic structure, including a brief coverage of energy sublevels and quantum numbers.

chapter three

General discussion of bonding (ionic and covalent):

Bonner, Francis T., Melba Phillips, and Jane Raymond, *Principles of Physical Science, Second Edition.* Reading, Mass.: Addison-Wesley Publishing Co., Inc., 1971. Chapter 19 contains a simple explanation, in depth, of the various types of bonds, electronegativity, and other concepts.

Epstein, H. T., *Elementary Biophysics.* Reading, Mass.: Addison-Wesley Publishing Co., Inc., 1963. Chapter 2 of this book contains a well-written, somewhat technical, coverage of the various types of forces involved in holding atoms and groups of atoms together. Good discussion of the topics of ionic and covalent bonding.

Drummond, A. H., *Atoms, Crystals and Molecules* (part 2). Columbus, Ohio: American Education Publications, 1964. This booklet is written especially for secondary school students. It contains a valuable discussion of ionic and covalent bonds, as well as hydrogen bonds and electronegativity. Chapters 3 through 7.

Ryschkewitsch, George E., *Chemical Bonding and the Geometry of Molecules.* New York: Reinhold Publishing Co., 1963. A paperback, this book is one of a series titled "Selected Topics in Modern Chemistry." It is probably of value only to the very interested or advanced student, but provides an excellent introduction to the concept of how bonds are formed and how they determine the actual geometric configuration of molecules.

Special types of bonds:

Lehninger, A. L., *Bioenergetics, Second Edition*. Menlo Park, Calif.: Benjamin, 1971. Chapters 4 and 5 deal specifically with ATP and the "high-energy" bond. This account is lucid, but requires some background and study.

Orgel, Leslie E. The Hydrogen Bond, in *Biophysical Science—A Study Program,* J. L. Oncley et al., editors. New York: Wiley, 1959, pp. 100–102. This brief article is a general survey of the nature of hydrogen bonds and the places they are found in the molecules of living systems.

chapter four

Van der Werf, Calvin, *Acids, Bases, and the Chemistry of the Covalent Bond*. New York: Reinhold Publishing Corp., 1962. This booklet is another in the series "Selected Topics in Modern Chemistry." It is one of the best single treatments of acids and bases. However, there is much more information in this book than may be useful to a beginning biology student, but the first three chapters will be especially helpful in elucidating certain features of the Brönsted-Lowry concept of acids and bases.

McElroy, W. D., *Cellular Physiology and Biochemistry*. Englewood Cliffs, N.J.: Prentice-Hall, Inc., 1961. This booklet is one in the series "Foundations of Modern Biology." In Chapter 3 there is a short discussion of pH, acids and bases (pp. 16–19). The general briefness of the treatment, however, makes this valuable only as a quick summary of the topic.

Chemical equilibrium:

Blum, Harold, *Time's Arrow and Evolution*. Princeton: Princeton Univ. Press, 1955. This is a provocative and well-written book. It is especially good on the concepts of free energy and entropy. The purpose of the book is to refute the argument that living organisms disobey the Second Law of Thermodynamics. Chapter 3 is an excellent discussion of chemical kinetics and energetics. The entire book gives a survey of much of the general information which has been covered in this chapter.

Bonner, Francis T., Melba Phillips, and Jane Raymond, *Principles of Physical Science,* Second Edition. Reading, Mass.: Addison-Wesley Publishing Co., Inc., 1971. In Chapter 21 is a discussion of chemical reversibility and equilibrium, and some of the mathematics involved in understanding equilibrium point.

Sisler, H. H., C. A. Van der Werf, and A. W. Davidson, *General Chemistry*. New York: Macmillan, 1959. Chapter 15 is devoted to a detailed discussion of equilibrium and the factors which influence it. This is a somewhat mathematical discussion.

Lehninger, A. L., *Bioenergetics* (Op. cit.). Chapters 1 and 2 deal with thermodynamics and the relation of the two laws of thermodynamics to chemical equilibrium and to overall biochemical processes.

Loewy, A. G. and P. Siekevitz, *Cell Structure and Function,* Second Edition. New York: Holt, Rinehart, and Winston, 1969. Chapter 2 of this book deals with "Life and the Second Law of Thermodynamics." It covers some points not discussed in the present book and provides a concise explanation of the law and its relation to living processes.

chapter nine

There is an almost endless amount of material written about proteins, much of it suitable for introductory students in biology. The list below contains only some of the most readable or the most useful advanced references of use in understanding modern biochemistry.

Anfinsen, Christian, *The Molecular Basis of Evolution.* New York: John Wiley and Sons, 1963. This really excellent book covers a wide variety of biochemical topics. Chapters 5 and 6 deal specifically with protein structure and the relation between biochemical activity and structure in these molecules. These two chapters (especially Chapter 5) are probably the best single discussion of protein outside of technical monographs.

Baldwin, Ernest, *The Nature of Biochemistry.* Cambridge: Cambridge Univ. Press, 1962. A well-written and not too complex book. Chapter 3 deals specifically with proteins and their macromolecular features. Chapter 4 deals with the physico-chemical behavior of proteins, while Chapter 5 discusses enzymes.

Kendrew, J. C., Three-dimensional Study of a Protein. *Scientific American,* December, 1961. An excellent discussion of the methods by which the three-dimensional structure of myoglobin was determined. There are some very helpful illustrations.

Kopple, K. D., *Peptides and Amino Acids.* New York: Benjamin, 1966. This is a valuable reference source on the structure and methods of studying proteins. It is considerably more detailed than the introductory student is likely to want, but can be used easily as a reference.

Loewy, A. G. and P. Siekevitz, *Cell Structure and Function, Second Edition.* New York: Holt, Rinehart and Winston, 1969. Chapter 8 deals with proteins, their structure and function.

Perutz, M., *Proteins and Nucleic Acids.* New York: Elsevier, 1962. Chapter 1, "The Structure of Proteins," is a very clear and readable account of how the three-dimensional structure of myoglobin was worked out. Perutz is one of the group of workers who carried out this task.

Stein, W. H. and S. Moore, The Chemical Structure of Proteins. *Scientific American,* February, 1961. The process by which the structure for the enzyme molecule ribonuclease was worked out is described.

chapter ten

Enzymes have been the subject of many short papers and lengthy monographs. Two of the most recent and best are available as paperbacks:

Bernhard, S., *The Structure and Function of Enzymes.* New York: Benjamin, 1968. Some parts of this book are highly technical and mathematical, but others are more readily comprehensible to the beginning biology student. Well illustrated, this book covers most of the fundamental aspects of enzyme structure, kinetics, and theories of molecular function.

Moss, D. W., *Enzymes.* London: Oliver and Boyd, 1968. This short paperback requires less mathematical and quantitative background than Bernhard's book listed above, yet it deals with all the major principles. It is less thorough on recent theories of enzyme control mechanisms such as the induced-fit hypothesis.

The standard work on enzymes is:

Dixon, M., and E. C. Webb, *Enzymes.* New York: Academic Press, 1964. This book is highly technical, but is one of the most complete reference sources on the principles of both enzyme structure and function, as well as information about specific enzyme systems. It is highly technical and mathematical.

chapter eleven

Like proteins, nucleic acids have been the subject of many recent publications. The list below contains only some of the less technical and more general sources about the nucleic acids and their role in coding specific protein structure.

Ingram, V. M., *Biosynthesis of Macromolecules,* Second Edition. Menlo Park, California: Benjamin, 1972. The most recent and up-to-date treatment of the mechanism of protein synthesis from DNA.

Lehninger, A. L., *Bioenergetics (Op. cit.),* Chapters 11 and 12. This is a good summary of protein synthesis and the role of the nucleic acids. Especially good is the author's treatment of the energetics involved.

Perutz, M., *Proteins and Nucleic Acids (Op. cit.).* Chapters 2 through 4 deal with the structure of the nucleic acids, and the translation of information into specific polypeptide structure.

Tables

TABLE OF AMINO ACIDS

Name (abbreviation)	Structural formula
Glycine (gly)	$H_3N^+—CH_2—COO^-$
Alanine (ala)	$H_3N^+—CH—COO^-$ CH_3
Valine (val)	$H_3N^+—CH—COO^-$ CH CH_3 CH_3
Leucine (leu)	$H_3N^+—CH—COO^-$ CH_2 CH CH_3 CH_3
Isoleucine (ileu)	$H_3N^+—CH—COO^-$ HC—CH_3 CH_2 CH_3
Phenylalanine (phe)	$H_3N^+—CH—COO^-$ CH_2 C HC CH HC CH C H
Tyrosine (tyr)	$H_3N^+—CH—COO^-$ CH_2 C HC CH HC CH C OH
Tryptophan (try)	$H_3N^+—CH—COO^-$ H CH_2 C C—C CH HC C CH N C H H
Aspartic acid (asp)	$H_3N^+—CH—COO^-$ CH_2 COO^-
Glutamic acid (glu)	$H_3N^+—CH—COO^-$ CH_2 CH_2 COO^-
Serine (ser)	$H_3N^+—CH—COO^-$ CH_2 OH
Threonine (thr)	$H_3N^+—CH—COO^-$ CH CH_3 OH
Cystine (cys)	$H_3N^+—CH—COO^-$ CH_2 S S CH_2 $H_3N^+—CH—COO^-$
Cysteine (cys H)	$H_3N^+—CH—COO^-$ CH_2 SH
Methionine (met)	$H_3N^+—CH—COO^-$ CH_2 CH_2 S CH_3
Arginine (arg)	$H_3N^+—CH—COO^-$ CH_2 CH_2 CH_2 NH C H_2N NH_2^+
Histidine (his)	$H_3N^+—CH—COO^-$ CH_2 C═CH HN NH^+ C H
Lysine (lys)	$H_3N^+—CH—COO^-$ CH_2 CH_2 CH_2 CH_2 NH_3^+

THE PERIODIC TABLE

Elements found in living organisms

- Always
- Variably

IA	IIA	IIIB	IVB	VB	VIB	VIIB	VIII			IB	IIB	IIIA	IVA	VA	VIA	VIIA	0
1 H 1.00797																	2 He 4.0026
3 Li 6.939	4 Be 9.012											5 B 10.811	6 C 12.011	7 N 14.007	8 O 15.9994	9 F 18.998	10 Ne 20.183
11 Na 22.990	12 Mg 24.312											13 Al 26.98	14 Si 28.086	15 P 30.97	16 S 32.064	17 Cl 35.453	18 Ar 39.95
19 K 39.102	20 Ca 40.08	21 Sc 44.96	22 Ti 47.90	23 V 50.94	24 Cr 52.00	25 Mn 54.94	26 Fe 55.85	27 Co 58.93	28 Ni 58.71	29 Cu 63.54	30 Zn 65.37	31 Ga 69.72	32 Ge 72.59	33 As 74.92	34 Se 78.96	35 Br 79.91	36 Kr 83.80
37 Rb 85.47	38 Sr 87.62	39 Y 88.91	40 Zr 91.22	41 Nb 92.91	42 Mo 95.94	43 Tc 99	44 Ru 101.07	45 Rh 102.91	46 Pd 106.4	47 Ag 107.87	48 Cd 112.40	49 In 114.82	50 Sn 118.69	51 Sb 121.75	52 Te 127.60	53 I 126.90	54 Xe 131.30
55 Cs 132.90	56 Ba 137.34	57–71 La series*	72 Hf 178.49	73 Ta 180.95	74 W 183.85	75 Re 186.2	76 Os 190.2	77 Ir 192.2	78 Pt 195.1	79 Au 196.97	80 Hg 200.59	81 Tl 204.37	82 Pb 207.19	83 Bi 208.98	84 Po 210	85 At 210	86 Rn 222
87 Fr 223	88 Ra 226	89– Ac series†															
*Lanthanide series		57 La 138.91	58 Ce 140.12	59 Pr 140.91	60 Nd 144.24	61 Pm 147	62 Sm 150.35	63 Eu 151.96	64 Gd 157.25	65 Tb 158.92	66 Dy 162.50	67 Ho 164.93	68 Er 167.26	69 Tm 168.93	70 Yb 173.04	71 Lu 174.97	
†Actinide series		89 Ac 227	90 Th 232.04	91 Pa 231	92 U 238.03	93 Np 237	94 Pu 239	95 Am 241	96 Cm 242	97 Bk 249	98 Cf 252	99 Es 254	100 Fm 253	101 Md	102 No	103 Lw	

In the periodic table, all known elements are arranged into groups which show similar chemical characteristics. These groups are represented by the vertical columns of elements. For example, all of the elements in the column under hydrogen have one valence electron and display certain common properties. All elements listed under beryllium (Be) have two outermost (valence) electrons and show certain specific characteristics in common. For each element, the atomic number is given above the symbol for the element, while the atomic weight is given below. The two rows of elements shown separately below the others are two series, the first of which fits between barium and hafnium, and the second of which follows radium. Elements in these series have special electron distributions which give them certain common properties. Hence, they best fit into the periodic table as separate groups. (From S. F. Peterson and R. G. Wymer, *Chemistry in Nuclear Technology*, Addison-Wesley Publishing Co., Inc., Reading, Mass., 1963.)

Index*

*Page numbers in boldface type refer to illustrations.